Universitext

Springer
*New York
Berlin
Heidelberg
Hong Kong
London
Milan
Paris
Tokyo*

Universitext

Editors (North America): S. Axler, F.W. Gehring, and K.A. Ribet

(continued after index)

An Invitation to Algebraic Geometry

Karen E. Smith, Lauri Kahanpää,
Pekka Kekäläinen and William Traves

Karen E. Smith
Department of Mathematics
University of Michigan
Ann Arbor, MI 48109-1109
USA

Lauri Kahanpää
Department of Mathematics
University of Jyvaeskylae
Jyvaeskylae, FIN-40351
Finland

Pekka Kekäläinen
Department of Computer Science and
    Applied Mathematics
University of Kuopio
70211 Kuopio
Finland

William Traves
US Naval Academy
Annapolis, MD 21402
USA

Mathematics Subject Classification (2000): 14-01

Library of Congress Cataloging-in-Publication Data
An invitation to algebraic geometry / Karen E. Smith ...[et al.].
    p.   cm.— (Universitext)
    Includes bibliographical references.
    ISBN 978-1-4419-3195-5
    1. Geometry, Algebraic.   I. Smith, Karen E., 1965–   II. Series.
QA564 .I62 2000
516.3′5—dc21   00-026595

Springer-Verlag is part of *Springer Science+Business Media*

springeronline.com

Notes for the Second Printing

The second printing of this book corrects the many typos and errors that were brought to our attention by readers from around the world. We have also added a few exercises and clarified parts of the text. We are grateful to all the readers who have helped improve our book, but owe particular thanks to Brian Conrad, Sándor Kovács, Grisha Stewart, and especially to Rahim Zaare Nahandi of the University of Tehran, who is engaged in translating this volume into Persian.

Karen E. Smith
Berkeley, CA, USA
March 2003

Preface

These notes grew out of a course at the University of Jyväskylä in January 1996 as part of Finland's new graduate school in mathematics. The course was suggested by Professor Kari Astala, who asked me to give a series of ten two-hour lectures entitled "Algebraic Geometry for Analysts." The audience consisted mainly of two groups of mathematicians: Ph.D. students from the Universities of Jyväskylä and Helsinki, and mature mathematicians whose research and training were quite far removed from algebra. Finland has a rich tradition in classical and topological analysis, and it was primarily in this tradition that my audience was educated, although there were representatives of another well–known Finnish school, mathematical logic.

I tried to conduct a course that would be accessible to everyone, but that would take participants beyond the standard course in algebraic geometry. I wanted to convey a feeling for the underlying algebraic principles of algebraic geometry. But equally important, I wanted to explain some of algebraic geometry's major achievements in the twentieth century, as well as some of the problems that occupy its practitioners today. With such ambitious goals, it was necessary to omit many proofs and sacrifice some rigor.

In light of the background of the audience, few algebraic prerequisites were presumed beyond a basic course in linear algebra. On the other hand, the language of elementary point-set topology and some basic facts from complex analysis were used freely, as was a passing familiarity with the definition of a manifold.

My sketchy lectures were beautifully written up and massaged into this text by Lauri Kahanpää and Pekka Kekäläinen. This was a Herculean effort, no less because of the excellent figures Lauri created with the computer. Extensive revisions to the Finnish text were carried out together with Lauri and Pekka; later Will Traves joined in to help with substantial revisions to the English version. What finally resulted is this book, and it would not have been possible without the valuable contributions of all members of our four-author team.

This book is intended for the working or the aspiring mathematician who is unfamiliar with algebraic geometry but wishes to gain an appreciation of its foundations and its goals with a minimum of prerequisites. It is not intended to compete with such comprehensive introductions as Hartshorne's or Shafarevich's texts, to which we freely refer for proofs and rigor. Rather, we hope that at least some readers will be inspired to undertake more serious study of this beautiful subject. This book is, in short, An Invitation to Algebraic Geometry.

Karen E. Smith
Jyväskylä, Finland
August 1998

Acknowledgments

The notes of Ari Lehtonen, Jouni Parkkonen, and Tero Kilpeläinen complemented those of authors Lauri and Pekka in producing a typed version of the original lectures. Comments of Osmo Pekonen, Ari Lehtonen, and Lassi Kurittu then helped eradicate most of the misprints and misunderstandings marring the first draft, and remarks of Bill Fulton later helped improve the manuscript. Artistic advice from Virpi Kauko greatly improved the pictures, although we were able to execute her suggestions only with the help of Ari Lehtonen's prize-winning Mathematica skills. Computer support from Ari and from Bonnie Freidman at MIT made working together feasible in Jyväskylä and in the US despite different computer systems. The suggestions of Manuel Blickle, Mario Bonk, Bill Fulton, Juha Heinonen, Eero Hyry, and Irena Swanson improved the final exposition. We are especially grateful to Eero for comments on the Finnish version; as one of the few algebraic geometers working in Finland, he advised us on the choices we made regarding mathematical terminology in the Finnish language. The Finnish craftwork of Liisa Heinonen provided instructive props for the lectures, most notably the traditional Christmas Blowup, whose image appears inside the cover of this book. Cooperation with Ari Lehtonen was crucial in creating the photograph. The lectures were hosted by the University of Jyväskylä mathematics department, and we are indebted to the chairman, Tapani Kuusalo, for making them possible. Finally, author Karen acknowledges the patience of her daughter Sanelma during the final stages of work on this project, and the support of her husband and babysitter, Juha Heinonen.

Karen E. Smith
Ann Arbor, Michigan, USA
July 2000

Contents

Index of Notation

$\mathbb{A}^n$ Affine n-space
$B_I(V)$ blow up of V along the ideal I
$B_p(V)$ blow up of V at the point p
$B_Y(V)$ blow up of V along the subcariety Y
$\mathbb{C}$ complex numbers
$\mathbb{C}[V]$ coordinate ring of the variety V
$\mathbb{C}(V)$ function field of V
dF differential of F
$\mathbb{F}_p$ field of p elements
$F^\#$ pull-back of a morphism F
$(\{F_i\})$ ideal generated by the polynomials F_i
$\mathbf{GL}(n, \mathbb{C})$ group of invertible $n \times n$ complex matrices
$\mathbf{Gr}(k, n)$ Grassmannian variety
Γ_F graph of the rational map F
$\mathbb{I}(V)$ ideal of functions vanishing on V
$\sqrt{I}$ radical of the ideal I
$|L|$ complete linear system
$\mathrm{maxSpec}(R)$ maximal spectrum of a ring R
$\mathfrak{M}_g$ moduli space of curves of genus g
$\mathcal{O}_V$ structure sheaf of V
Ω_X sheaf of sections of the cotangent bundle
ω_X canonical line bundle
$\mathbb{P}^n$ Projective n-space
$\check{\mathbb{P}}^n$ dual projective n-space
$[a_0 : \cdots : a_n]$ point in $\mathbb{P}^n$

$\mathbf{PGL}(n, \mathbb{C})$ automorphism group of $\mathbb{P}^{n-1}$

$\mathbb{R}$ real numbers

$\tilde{R}$ sheaf associated to $\mathrm{Spec}(R)$

$\mathcal{R}(U)$ sections of a sheaf $\mathcal{R}$ over an open set U

$X \dashrightarrow Y$ rational map from X to Y

Sec X secant variety to X

$\mathbf{SL}(n, \mathbb{C})$ group of $n \times n$ complex matrices with determinant 1

$\mathrm{Spec}(R)$ spectrum of a ring R

$\Sigma_{m,n}$ Segre mapping

Sing V singular locus of V

Tan X tangent variety to X

$T_p V$ tangent space to V at the point p

TV total tangent bundle to V

Θ_X sheaf of sections of the tangent bundle

$\mathbf{U}(n)$ group of unitary $(n \times n)$-matrices

$\bar{V}$ projective closure of V

$\mathbb{V}(\{F_i\})$ common zeros of the polynomials F_i

ν_d Veronese mapping of degree d

$\mathbb{Z}$ integers

1
Affine Algebraic Varieties

Algebraic geometers study zero loci of polynomials. More accurately, they study geometric objects, called algebraic varieties, that can be described locally as zero loci of polynomials. For example, every high school mathematics student has studied a bit of algebraic geometry, in learning the basic properties of conic sections such as parabolas and hyperbolas.

Algebraic geometry is a thriving discipline with a rich history. In ancient Greece, mathematicians such as Apollonius probably knew that a non-degenerate plane conic is uniquely determined by five tangent lines, a problem that would cause many modern students of algebraic geometry to pause. But it was not until the introduction of the Cartesian coordinate system in the seventeenth century, when it became possible to study conic sections by considering quadratic polynomials, that the subject of algebraic geometry could really take off.

By the mid-nineteenth century, algebraic geometry was flourishing. On the one hand, Riemann realized that compact Riemann surfaces can always be described by polynomial equations. On the other hand, particular examples of algebraic varieties, such as quadric and cubic surfaces (zero loci of a single quadratic or cubic polynomial in three variables) were well known and intensely studied. For example, it was understood that every quadric surface is perfectly covered by a family of disjoint lines, whereas every cubic surface contains exactly twenty-seven lines. Detailed studies of the ways in which these twenty-seven lines can be configured and how they can vary in families occupied the attention of numerous nineteenth-century mathematicians.

The remarkable intuition of the turn-of-the-century algebraic geometers eventually began to falter as the subject grew beyond its somewhat shaky logical foundations. Led by David Hilbert, mathematical culture shifted toward a greater emphasis on rigor, and soon algebraic geometry fell out of favor as gaps and even some errors appeared in the subject. Luckily, the spirit and techniques of algebraic geometry were kept alive, primarily by Italian mathematicians. By the mid-twentieth century, with the efforts of mathematicians such as David Hilbert and Emmy Noether, algebra was sufficiently developed so as to be able once again to support this beautiful and important subject.

In the middle of the twentieth century, Oscar Zariski and André Weil spent a good portion of their careers redeveloping the foundations of algebraic geometry on firm mathematical ground. This was not a mere process of filling in details left unstated before, but a revolutionary new approach, based on analyzing the algebraic properties of the set of all polynomial functions on an algebraic variety. These innovations revealed deep connections between previously separate areas of mathematics, such as number theory and the theory of Riemann surfaces, and eventually allowed Alexander Grothendieck to carry algebraic geometry to dizzying heights of abstraction in the last half of the century. This abstraction has simplified, unified, and greatly advanced the subject, and has provided powerful tools used to solve difficult problems. Today, algebraic geometry touches nearly every branch of mathematics.

An unfortunate effect of this late-twentieth-century abstraction is that it has sometimes made algebraic geometry appear impenetrable to outsiders. Nonetheless, as we hope to convey in this *Invitation to Algebraic Geometry*, the main objects of study in algebraic geometry, affine and projective algebraic varieties, and the main research questions about them, are as interesting and accessible as ever.

1.1   Definition and Examples

An algebraic variety is a geometric object that locally resembles the zero locus of a collection of polynomials. The idea of "locally resembling" is familiar to those who have studied manifolds, which are geometric objects locally resembling Euclidean space. We begin our study of algebraic geometry by considering this local picture in detail, the study of affine algebraic varieties.

Definition: An *affine algebraic variety* is the common zero set of a collection $\{F_i\}_{i \in I}$ of complex polynomials on complex n-space $\mathbb{C}^n$. We write

$$V = \mathbb{V}(\{F_i\}_{i \in I}) \subset \mathbb{C}^n$$

for this set of common zeros. Note that the indexing set I can be arbitrary, not necessarily finite or even countable.

For example, $V = \mathbb{V}(x_1, x_2) \subset \mathbb{C}^3$ is the complex line in $\mathbb{C}^3$ consisting of the x_3-axis.

This definition of an affine algebraic variety should be considered only a working preliminary definition. The problem is that it depends on considerations extrinsic to the objects themselves, namely the embedding of the affine variety in the particular affine space $\mathbb{C}^n$. Later, in Section 4.1, we will refine and expand our definition of an affine algebraic variety in order to make it a more intrinsic notion.

Strictly speaking, what we have defined above should be called a complex affine algebraic variety, because we are considering our varieties over the complex numbers. The field of complex numbers may be replaced by any other field, such as the field $\mathbb{R}$ of real numbers, the field $\mathbb{Q}$ of rational numbers, or even a finite field. For reasons we will see later, using complex numbers instead of real numbers makes algebraic geometry easier, and in order to keep this book as close as possible to familiar territory, we will work only over the complex numbers $\mathbb{C}$. However, the reader should bear in mind the possibility of using different fields; this flexibility allows algebraic geometry to be applied to problems in number theory (by using the rational numbers or some p-adic fields).

Examples:

(1) The space $\mathbb{C}^n$; the empty set; and one-point sets, *singletons*, are trivial examples of affine algebraic varieties:

$$\begin{aligned} \mathbb{C}^n &= \mathbb{V}(0); \\ \emptyset &= \mathbb{V}(1); \\ \{(a_1, \ldots, a_n)\} &= \mathbb{V}(x_1 - a_1, \ldots, x_n - a_n). \end{aligned}$$

We call the space $\mathbb{C}$ the *complex line*, and the space $\mathbb{C}^2$ the *complex plane*. Confusing as it may seem, the complex line $\mathbb{C}$ is called the "complex plane" in some other branches of mathematics. In general, the space $\mathbb{C}^n$ is called complex n-space or affine n-space.

When drawing a sketch of an affine algebraic variety V we will, of course, draw only its *real points* $V \cap \mathbb{R}^n$.

(2) An *affine plane curve* is the zero set of one complex polynomial in the complex plane $\mathbb{C}^2$. Figures 1.1, 1.2, and 1.3 show examples of plane curves.

(3) The zero set of a single polynomial in arbitrary dimension is called a *hypersurface* in $\mathbb{C}^n$. The quadratic cone in Figure 1.4 is a typical example of a hypersurface.

(4) The zero set of a linear (degree-one) polynomial is an affine algebraic variety called an affine *hyperplane*. For example, the line defined by $ax + by = c$ is a hyperplane in the complex plane $\mathbb{C}^2$, where here a, b, and c are complex scalars. A *linear affine algebraic variety* is the common zero set of

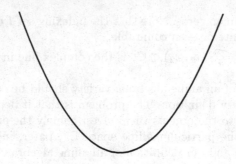

Figure 1.1. $\mathbb{V}(y - x^2) \subset \mathbb{C}^2$

Figure 1.2. $\mathbb{V}(x^2 y + xy^2 - x^4 - y^4) \subset \mathbb{C}^2$

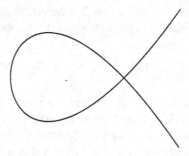

Figure 1.3. $\mathbb{V}(y^2 - x^2 - x^3) \subset \mathbb{C}^2$

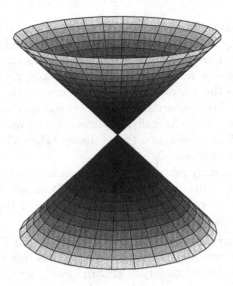

Figure 1.4. The quadratic cone $\mathbb{V}(x^2 + y^2 - z^2)$ in $\mathbb{C}^3$

a collection of linear polynomials of the form

$$a_1 x_1 + a_2 x_2 + \ldots + a_n x_n - b$$

in $\mathbb{C}^n$. If there are k linearly independent polynomials, the linear variety is a complex space of dimension $n - k$.

(5) The set of all $n \times n$ matrices can be identified with the set $\mathbb{C}^{n^2}$. This space contains some familiar objects as affine algebraic varieties. For instance, the subset $\mathbf{SL}(n, \mathbb{C})$ of matrices of determinant 1 forms an affine algebraic variety in $\mathbb{C}^{n^2}$, the hypersurface defined by the polynomial $\Delta - 1$, where Δ denotes the *determinant*

$$\Delta(x_{ij}) = \det \begin{bmatrix} x_{11} & \cdots & x_{1n} \\ \vdots & & \vdots \\ x_{n1} & \cdots & x_{nn} \end{bmatrix},$$

which is obviously a polynomial in the n^2 variables x_{ij}.

(6) A *determinantal variety* is the set in $\mathbb{C}^{n^2}$ of all matrices of rank at most k, where k is some fixed natural number. For $k \geq n$ the determinantal variety is the whole space $\mathbb{C}^{n^2}$, but for $k < n$ the rank of a matrix A is at most k if and only if all its $(k + 1) \times (k + 1)$ subdeterminants vanish. Because the subdeterminants are polynomials in the variables x_{ij}, the set of matrices of rank at most k is an affine algebraic variety.

Nonexamples:

(1) An open ball in the usual Euclidean topology on $\mathbb{C}^n$ is not an algebraic variety. In fact, every affine algebraic variety in $\mathbb{C}^n$ is closed in the Euclidean topology, as we will show in Exercise 1.1.1. For this reason, the set $\mathbf{GL}(n, \mathbb{C})$ of invertible matrices is not an affine algebraic variety as so far defined. Indeed, $\mathbf{GL}(n, \mathbb{C})$ is the complement of the algebraic variety in $\mathbb{C}^{n^2}$ defined by the vanishing of the determinant polynomial, and so is open in the Eucliean topology on $\mathbb{C}^{n^2}$. Actually, we later expand our definition of an affine algebraic variety in Section 4.1, and the set $\mathbf{GL}(n, \mathbb{C})$ will be an affine variety in this expanded sense.

The set $\mathbf{U}(n)$ of unitary matrices is not a complex algebraic variety, even under our expanded definition. Recall that an $n \times n$ matrix with complex entries is unitary if its columns are orthonormal under the complex inner product $\langle z, w \rangle = z \cdot \overline{w}^t = \sum_{i=1}^{n} z_i \overline{w}_i$.

(2) The closed square $\{(x, y) \in \mathbb{C}^2 : |x| \le 1, |y| \le 1\}$ in $\mathbb{C}^2$ is an example of a closed set that is not an algebraic variety. This follows from the fact that no nontrivial algebraic variety in $\mathbb{C}^2$ can have interior points, since the zero set of one nonzero polynomial has no interior points.

(3) Graphs of transcendental functions are not algebraic varieties. For example, the zero set of the function $y - e^x$ is not an algebraic variety. See Exercise 6 in Section 2.3.

Exercise 1.1.1. Show that every affine algebraic variety in $\mathbb{C}^n$ is closed in the Euclidean topology. (Hint: Polynomials are continuous functions from $\mathbb{C}^n$ to $\mathbb{C}$, so their zero sets are closed.)

Exercise 1.1.2. A subvariety of an affine algebraic variety $V \subset \mathbb{C}^n$ is an affine algebraic variety $W \subset \mathbb{C}^n$ that is contained in V. Show that the set $\mathbf{U}(n)$ is not an affine algebraic subvariety of $\mathbb{C}^{n^2}$. Show, however, that it can be described as the zero locus of a collection of polynomials with real coefficients in $\mathbb{R}^{2n^2}$, that is, it is a *real algebraic variety*.

1.2   The Zariski Topology

The intersection of any number of affine algebraic varieties in $\mathbb{C}^n$ is an affine algebraic variety. Indeed, the intersection is defined by the union of the sets of polynomials defining the given varieties. For example, an intersection of two algebraic varieties can be written

$$\mathbb{V}(\{F_i\}_{i \in I}) \cap \mathbb{V}(\{F_j\}_{j \in J}) = \mathbb{V}(\{F_i\}_{i \in I \cup J}).$$

The *twisted cubic curve* pictured in Figure 1.5 offers a concrete example of an intersection of two surfaces.

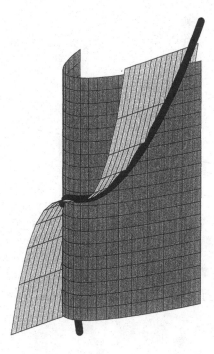

Figure 1.5. $V = \mathbb{V}(x^2 - y, x^3 - z) = \mathbb{V}(x^2 - y) \cap \mathbb{V}(x^3 - z)$.

The union of two affine algebraic varieties in $\mathbb{C}^n$ is an affine algebraic variety. For example, it is easy to see that the union of two hypersurfaces is defined by the product of the corresponding polynomials:

$$\mathbb{V}(F_1) \cup \mathbb{V}(F_2) = \mathbb{V}(F_1 F_2).$$

Indeed, the polynomial $F_1 F_2$ vanishes at a point p if and only if one (or both) of F_1 or F_2 vanishes at p. For example, the union of the x and y-axes in the plane is the zero set of the single polynomial xy.

More generally, the union of two arbitrary affine algebraic varieties is defined by the set of all pairwise products of the polynomials defining the original varieties:

$$\mathbb{V}(\{F_i\}_{i \in I}) \cup \mathbb{V}(\{F_i\}_{i \in J}) = \mathbb{V}(\{F_i F_j\}_{(i,j) \in I \times J}).$$

For example, Figure 1.6 depicts the union of the yz-plane (defined by the vanishing of x) and the x-axis (defined by the vanishing of both y and z). This union is the common vanishing set of the polynomials xy and xz.

We have verified that the empty set, the whole space $\mathbb{C}^n$, the intersection of arbitrarily many affine algebraic varieties, and the union of two (and, by induction, finitely many) affine algebraic varieties are all affine algebraic varieties in $\mathbb{C}^n$. Therefore, the set $\mathcal{Z}$ of all complements of affine algebraic

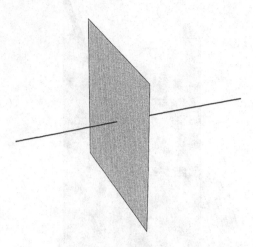

Figure 1.6. $V = \mathbb{V}(y, z) \cup \mathbb{V}(x) = \mathbb{V}(xy, xz)$

sets satisfies the four axioms defining a *topology* in $\mathbb{C}^n$: The whole space and the empty set are in $\mathcal{Z}$, as well as the intersection of finitely many elements of $\mathcal{Z}$ and the union of arbitrarily many elements of $\mathcal{Z}$. So, $\mathcal{Z}$ turns $\mathbb{C}^n$ into a topological space, where the open sets are exactly the complements of affine algebraic varieties. This topology is called the *Zariski topology* on $\mathbb{C}^n$. To emphasize the difference between the vector space $\mathbb{C}^n$ and the set $\mathbb{C}^n$ considered as a topological space with its Zariski topology, we will denote the topological space by $\mathbb{A}^n$, and call it *affine n-space*. In particular, there is no distinguished "origin" in $\mathbb{A}^n$. Despite this, we often implicitly choose coordinates and refer to "the origin in $\mathbb{A}^n$."

Because every affine algebraic variety is closed in the Euclidean topology, every Zariski-closed set is closed in the Euclidean topology. The converse, however, is false; the Zariski topology is much coarser than the Euclidean topology on $\mathbb{C}^n$. The Euclidean topology has a basis consisting of open balls of arbitrarily small radius; in contrast, the nonempty Zariski-open sets are very large, like the complements of curves or surfaces in 3-space. Every nonempty Zariski-open set is dense in both the Zariski topology and in the Euclidean topology, so in particular, no Zariski open set is bounded in the usual Euclidean topology. The intersection of two nonempty Zariski-open sets of $\mathbb{A}^n$ is never empty, so the Zariski topology cannot be a Hausdorff topology. A set may well be Zariski-compact without being Zariski-closed or even closed in the Euclidean topology.[1] See Exercise 2.3.5.

In contrast to the Euclidean topology, the Zariski topology also makes sense when we are dealing with fields other than the complex numbers. If we wish to study zero sets of polynomials in the space $\mathbb{K}^n$, where $\mathbb{K}$ is an arbitrary field, the Zariski topology is at our disposal, though the Euclidean topology is not.

Every affine algebraic variety inherits a topology from the ambient space $\mathbb{A}^n$. The Zariski topology on an affine algebraic variety V is the subspace topology on V induced by the Zariski topology of $\mathbb{A}^n$. In particular, the closed sets in V will be the intersections $V \cap W$ of V with affine algebraic varieties $W \subset \mathbb{A}^n$. In other words, the closed sets of V are the *affine algebraic subvarieties* of V.

Example: All proper Zariski-closed sets of the parabola $V = \mathbb{V}(y - x^2) \subset \mathbb{A}^2$ are finite. Indeed, the Zariski topology on any plane curve is the *cofinite topology*, provided that the curve is not a union of two other curves.

In algebraic geometry, varieties are considered with their Zariski topology. Unless otherwise stated, topological concepts in this book will always refer to the Zariski topology.

Exercise 1.2.1. Show that the union of two affine algebraic varieties in complex n-space is an affine algebraic variety.

Exercise 1.2.2. Show that the Zariski topology on $\mathbb{A}^2$ is not the product topology on $\mathbb{A}^1 \times \mathbb{A}^1$. (Hint: Consider the diagonal.)

Exercise 1.2.3. Show that the twisted cubic curve depicted in Figure 1.5 consists of all points in $\mathbb{A}^3$ of the form (t, t^2, t^3), where $t \in \mathbb{C}$.

1.3 Morphisms of Affine Algebraic Varieties

Just as an algebraic variety is given by polynomials, a morphism of algebraic varieties is also given by polynomials.

The simplest example of a morphism of algebraic varieties is a polynomial map

$$\mathbb{A}^n \xrightarrow{F} \mathbb{A}^m,$$
$$x \longmapsto (F_1(x), F_2(x), \ldots, F_m(x))$$

---

[1] A *compact* space is a topological space for which every open cover has a finite subcover. Some authors call these spaces *quasicompact*, reserving the term "compact" for Hausdorff spaces with this property.

where by *polynomial map* we mean that each of the components F_i of F is a polynomial in the n coordinates $x_1, \ldots, x_n$ of $\mathbb{A}^n$. In general, a morphism of affine algebraic varieties is defined as follows.

Definition: Let $V \subseteq \mathbb{A}^n$ and $W \subseteq \mathbb{A}^m$ be affine algebraic varieties. A map $V \xrightarrow{F} W$ is a *morphism of algebraic varieties* if it is the restriction of a polynomial map on the ambient affine spaces $\mathbb{A}^n \to \mathbb{A}^m$.

A morphism of algebraic varieties $V \to W$ is an *isomorphism* if it admits an inverse morphism, that is, if it is bijective and its inverse is also a morphism. We will say that two affine algebraic varieties are *isomorphic* if there exists an isomorphism between them.

Example: An affine change of coordinates of $\mathbb{A}^n$ is an example of an isomorphism of $\mathbb{A}^n$ with itself, or an *automorphism*. Explicitly, let

$$L_i(x) = \lambda_{i1}x_1 + \ldots + \lambda_{in}x_n + \mu_i$$

be a degree one polynomial in $x_1, \ldots, x_n$, where each λ_{ij} and μ_i is in $\mathbb{C}$. Then the the map

$$\begin{aligned} \mathbb{A}^n &\longrightarrow \mathbb{A}^n, \\ x &\longmapsto (L_1(x), \ldots, L_n(x)), \end{aligned}$$

is a morphism of algebraic varieties. It is an isomorphism if and only if the matrix (λ_{ij}) is invertible.

Example: The projection $\mathbb{A}^2 \to \mathbb{A}^1$ sending (x, y) to x is a morphism of algebraic varieties. It cannot be an isomorphism because it is not bijective.

Example: Let C be the plane parabola defined by the vanishing of the polynomial $y - x^2$. The morphism

$$\begin{aligned} \mathbb{A}^1 &\longrightarrow C, \\ t &\longmapsto (t, t^2), \end{aligned}$$

is easily seen to be an isomorphism. Its inverse map is given by the (restriction of) the projection

$$\begin{aligned} \mathbb{A}^2 \supset C &\longrightarrow \mathbb{A}^1, \\ (x, y) &\longmapsto x. \end{aligned}$$

This correspondence is indicated in Figure 1.7.

The other projection $(x, y) \mapsto y$ defines a two-to-one morphism from the parabola C to the affine line.

It is important to realize that a morphism of algebraic varieties need not send subvarieties to subvarieties, that is, a morphism need not be a closed

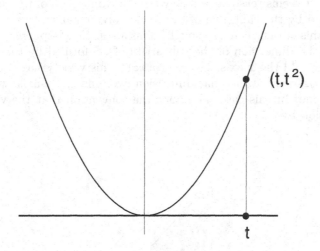

Figure 1.7. The parabola is isomorphic to the line

map. A simple example is given by the projection

$$\mathbb{A}^2 \xrightarrow{\pi} \mathbb{A}^1,$$
$$(x,y) \longmapsto x.$$

The hyperbola $\mathbb{V}(xy - 1) = \{(t, t^{-1}) \mid t \neq 0\}$ is a closed set of $\mathbb{A}^2$ that is mapped onto the set $\mathbb{A}^1 \smallsetminus \{0\}$, which is not a Zariski-closed subset of $\mathbb{A}^1$.

Exercise 1.3.1. Let $V \xrightarrow{F} W$ be a morphism of affine algebraic varieties. Prove that F is continuous in the Zariski topology.

Exercise 1.3.2. Show that the twisted cubic V of Figure 1.5 is isomorphic to the affine line by constructing an explicit isomorphism $\mathbb{A}^1 \to V$. (Hint: See Exercise 1.2.3)

1.4 Dimension

Developing a good theory of dimension is a challenging problem in any branch of mathematics, and algebraic geometry is no exception. On the other hand, most readers already have some feeling for what we mean by the dimension of an algebraic variety. To develop the subject carefully, it is best to take a more algebraic approach. Here, we will simply define and discuss the basic facts about dimension, relying on the reader's intuition and referring to [37, Chapter I, Section 6] for technical details.

First, a basic example: The affine n-space $\mathbb{A}^n$ has dimension n.

Likewise, it seems reasonable to say that the dimension of the subvariety of $\mathbb{A}^3$ defined by the vanishing of the single polynomial $x^2 + y^2 + z^2 - 1$ is two, since this variety can be thought of as a complex 2-sphere.

What is the dimension of the subvariety of $\mathbb{A}^3$ formed by the union of the yz-plane and the x-axis, $V = \mathbb{V}(xy, xz)$? This variety has two *components*: The yz-plane, which has dimension two, and the x-axis, which has dimension one. In this case, we adopt the convention that the variety V has dimension two.

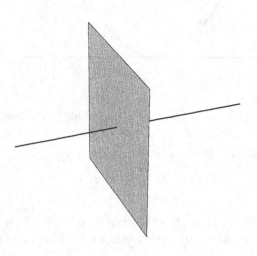

Figure 1.8. A variety with two components

Varieties which cannot be written as the nontrivial union of two subvarieties are said to be *irreducible*. The yz-plane and the x-axis are the irreducible components of the variety V above.

We now define the dimension of an algebraic variety precisely. The *dimension* $\dim V$ of a variety V is defined to be the length d of the longest possible chain of distinct nonempty irreducible subvarieties of V,

$$V_d \supsetneq V_{d-1} \supsetneq \cdots \supsetneq V_1 \supsetneq V_0.$$

In particular, the line $\mathbb{A}^1$ has dimension 1 because its only proper irreducible subvarieties are singletons: {line} $\supsetneq$ {point}. With this definition, the dimension of a variety is the same as the maximum dimension of its irreducible components. A variety is said to be *equidimensional* if all of its irreducible components have the same dimension. The variety in Figure 1.8 is not equidimensional.

The *codimension* of the algebraic variety $V \subset \mathbb{A}^n$ is the number $\mathrm{codim} V = n - \dim V$. Of course, the codimension depends on the ambient space: A line in the plane has codimension one, whereas a line in 3-space has codimension two.

We can also speak of the *dimension* $\dim_x(V)$ *of V near a point x* in V. This is just the length of the longest possible chain of irreducible subvarieties ending with $\{x\}$,

$$V_d \supsetneq V_{d-1} \supsetneq \cdots \supsetneq V_1 \supsetneq V_0 = \{x\}.$$

Note that

$$\dim V = \sup\{\dim_x V : x \in V\}$$

It can be proved that the dimension of an irreducible variety is the same at all points.

Admittedly, it is not obvious from our definition of dimension that $\mathbb{A}^n$ is n-dimensional. However, it is easy to verify that the dimension of $\mathbb{A}^n$ is at least n, by considering an increasing chain of linear subvarieties. To show that the dimension of $\mathbb{A}^n$ is exactly n we would need to develop more algebraic machinery; see [37]. We will at least show in Section 2.3 that the dimension of any variety, and $\mathbb{A}^n$ in particular, is finite. Given that the dimension of $\mathbb{A}^n$ is n, it is of course clear that the dimension of any proper subvariety of $\mathbb{A}^n$ is at most $n - 1$.

Our definition of dimension is compatible with the concept of dimension for manifolds: It turns out that every variety contains a dense Zariski-open subset of "smooth points," where the variety admits the structure of a complex manifold. At such points, our definition of dimension agrees with the dimension as a complex manifold. A proof can be found in the book by Shafarevich [37, Book I, Section 6, Theorem 1, page 54], where the reader will also find a more rigorous development of dimension theory for algebraic varieties.

Exercise 1.4.1. Show that dimension is an invariant of the isomorphism class of a variety. That is, affine algebraic varieties that are isomorphic to each other have the same dimension.

Exercise 1.4.2. Show that if $X \to Y$ is a surjective morphism of affine algebraic varieties, then the dimension of X is at least as large as the dimension of Y.

Exercise 1.4.3. Show that a hypersurface in $\mathbb{A}^n$ is irreducible if and only if the defining equation F is a power of an irreducible polynomial G (that is, G cannot be written as a product of two nonconstant polynomials).

2

Algebraic Foundations

2.1 A Quick Review of Commutative Ring Theory

Much of the power and rigor of algebraic geometry comes from the fact that geometric questions can be translated into purely algebraic problems.

Consider the set $\mathbb{C}[x_1, \ldots, x_n]$ of all complex polynomial functions in n variables. Because the sum of two polynomials is a polynomial and the product of two polynomials is a polynomial, this set forms a commutative ring in a natural way; the constant polynomial 1 is the multiplicative identity and the constant polynomial 0 is the additive identity. Indeed, because the ring $\mathbb{C}[x_1, \ldots, x_n]$ contains the constant polynomial functions, this polynomial ring naturally forms a $\mathbb{C}$-algebra, that is, it is a (commutative) ring containing $\mathbb{C}$ as a subring.

There is a surprisingly close relation between the study of the algebraic variety $\mathbb{A}^n$ and the study of the ring $\mathbb{C}[x_1, \ldots, x_n]$ of polynomial functions on it. As we will soon see, the subvarieties of $\mathbb{A}^n$ correspond precisely to certain kinds of ideals in the ring $\mathbb{C}[x_1, \ldots, x_n]$.

We now quickly recall some basic definitions and facts from algebra that will be needed in the coming sections. The reader is encouraged to skim over this section quickly and return to it as necessary.

For us, a *ring* will always be associative, commutative and contain a multiplicative unit 1. A mapping $R \xrightarrow{f} S$ between rings is called a *ring homomorphism* or just a *ring map* if it preserves sums, products, and the unit. A nonempty subset $I \subset R$ of a ring R is an *ideal* if it is closed under addition and under multiplication by elements of R. The *trivial* ideals are the zero ideal $\{0\}$ and the unit ideal R.

Example: The set of all polynomials with zero constant term is an ideal in the polynomial ring $R = \mathbb{C}[x, y]$.

An intersection of arbitrarily many ideals is also an ideal. So it makes sense to talk about the ideal *generated by* a set $J \subset R$. This is just the ideal

$$(J) = \bigcap \{ I \mid J \subset I, I \subset R \text{ an ideal } \}.$$

From this, it is clear that the ideal (J) is the smallest ideal containing the set J. We can also think of the ideal (J) generated by a set $J \subset R$ as the collection of all finite R-linear combinations of elements in J, that is, all elements of the form $r_1 j_1 + \cdots + r_n j_n$ where $r_i \in R$ and $j_i \in J$.

An ideal I is said to be *finitely generated* if there is a finite set $J = \{j_1, \ldots, j_n\} \subset R$ generating I. In this case we write $I = (J) = (j_1, \ldots, j_n)$.

Example: The elements of the ideal $I \subset \mathbb{C}[x, y]$ of polynomials with zero constant term are of the form $xP(x, y) + yQ(x, y)$, where $P, Q \in \mathbb{C}[x, y]$. Thus the ideal I is generated by the polynomials x and y; we write $I = (x, y)$.

The preimage of any ideal under a ring map is an ideal. In particular, the *kernel* $f^{-1}(\{0\})$ of a ring map $R \xrightarrow{f} S$ is an ideal.

There are some particularly important types of ideals:

- An ideal $\mathfrak{m} \subsetneq R$ is *maximal* if the only ideal strictly containing it is the unit ideal R.

- An ideal $\mathfrak{p} \subsetneq R$ is called *prime* if $fg \in \mathfrak{p}$ only when $f \in \mathfrak{p}$ or $g \in \mathfrak{p}$.

- An ideal $I \subset R$ is called *radical* if it is equal to its radical, where the radical of I is defined to be

$$\sqrt{I} := \{ f \in R \mid f^n \in I \text{ for some } n > 0 \}.$$

If $I \subset R$ is an ideal, then the set of cosets $R/I = \{ [x] = x + I \mid x \in R \}$ forms a ring with the natural operations $[x] + [y] = [x+y]$ and $[x][y] = [xy]$. There is a *canonical surjection* $R \to R/I$ sending an element x to the corresponding coset $[x]$. The kernel of this map is the ideal I.

Because the canonical surjection $R \xrightarrow{\pi} R/I$ is a homomorphism, the preimage $\pi^{-1}(J)$ of any ideal $J \subset R/I$ is an ideal of R containing I. On the other hand, π also maps every ideal $K \subset R$ containing I onto an ideal in the quotient ring. Therefore, the ideals of the quotient ring stand in a one-to-one correspondence with the ideals of R that contain I:

$$\{\text{ideals in } R/I\} \longleftrightarrow \{\text{ideals in } R \text{ containing } I\}.$$

This bijection carries maximal (respectively prime, radical) ideals to maximal (respectively prime, radical) ideals.

In this book, nearly all the rings we consider will be $\mathbb{C}$-algebras. Recall that a ring R is called a $\mathbb{C}$-algebra if it contains $\mathbb{C}$ as a subring. Every $\mathbb{C}$-algebra is also a $\mathbb{C}$-vector space, where the addition of vectors is defined by the addition in R and the multiplication of a scalar $\lambda \in \mathbb{C}$ and a vector $r \in R$ is defined by the multiplication in R.

We can define concepts for $\mathbb{C}$-algebras that are analogous to those for rings and ideals:

- The $\mathbb{C}$-subalgebra *generated* by a subset J of a $\mathbb{C}$-algebra R is

$$\bigcap \{A \mid J \subset A, A \subset R \text{ a } \mathbb{C}\text{-subalgebra}\}.$$

  This is the smallest $\mathbb{C}$-subalgebra containing J. The $\mathbb{C}$-subalgebra generated by J consists of all elements of R that can be written as *polynomials* in the elements of J with coefficients in $\mathbb{C}$.

- The algebra R is *finitely generated* if it is generated by some finite set $J \subset R$. For example, the polynomial ring $\mathbb{C}[x, y]$ is a $\mathbb{C}$-algebra, because it contains the subring $\mathbb{C}$ of constant functions. It is finitely generated as a $\mathbb{C}$-algebra by the elements x and y.

- If R and S are $\mathbb{C}$-algebras, then a map

$$R \xrightarrow{\phi} S$$

  is said to be a $\mathbb{C}$-*algebra homomorphism* if it is a ring map and if it is linear over $\mathbb{C}$, that is, $\phi(\lambda r) = \lambda \phi(r)$ for all $\lambda \in \mathbb{C}$ and $r \in R$.

An example of a ring map that is not a $\mathbb{C}$-algebra map is the complex conjugation map

$$\mathbb{C}[x] \longrightarrow \mathbb{C}[x],$$
$$a_0 + a_1 x + \ldots + a_n x^n \longmapsto \overline{a_0} + \overline{a_1} x + \ldots + \overline{a_n} x^n,$$

although this does define an $\mathbb{R}$-linear map.

Any $\mathbb{C}$-algebra map is determined by the images of any set of $\mathbb{C}$-algebra generators. For example, a $\mathbb{C}$-algebra map

$$\frac{\mathbb{C}[x, y]}{(x^2 + y^3)} \xrightarrow{\phi} \mathbb{C}[z]$$

is completely determined by the images of the generators x and y of $\frac{\mathbb{C}[x,y]}{(x^2+y^3)}$. For instance, ϕ would be determined by the data $\phi(x) = z^3$ and $\phi(y) = -z^2$. Note that the images of the $\mathbb{C}$-algebra generators cannot be arbitrary: The images must satisfy the same relations satisfied by the generators.

Exercise 2.1.1. Prove that every maximal ideal is prime, and every prime ideal is radical. Also prove that the radical $\sqrt{I}$ of an ideal I is an ideal.

Exercise 2.1.2. Prove that an ideal $\mathfrak{m}$ is maximal if and only if $R/\mathfrak{m}$ is a field. Prove that an ideal P is prime if and only if the ring R/P is a *domain*, that is, R/P has the property that whenever $xy = 0$, either $x = 0$ or $y = 0$.

Exercise 2.1.3. Let $I \subset S$ be any ideal. Prove that any ring map $\sigma : R \to S$ induces an injective homomorphism of rings: $R/\sigma^{-1}(I) \to S/I$. Conclude that if I is prime, so is $\sigma^{-1}(I)$.

Exercise 2.1.4. A ring R is *reduced* if for all $f \in R$ and each $n \in \mathbb{N}$,

$$f^n = 0 \iff f = 0.$$

That is, R is reduced if it contains no nonzero *nilpotent* elements. Prove that a ring R is reduced if and only if the zero ideal is radical.

Exercise 2.1.5. Prove that the quotient ring R/I is reduced if and only if I is a radical ideal.

Exercise 2.1.6. Let R be a $\mathbb{C}$-algebra, and let I be an ideal of R. Prove that the natural surjection $R \to R/I$ is a $\mathbb{C}$-algebra map.

2.2   Hilbert's Basis Theorem

Although the definition allows arbitrarily many polynomials, it turns out that every affine algebraic variety is the common zero set of finitely many polynomials. This follows from the important *Noetherian* property of polynomial rings.

Definition: A ring R is *Noetherian* if all its ideals are finitely generated.

Hilbert's Basis Theorem: If a ring R is Noetherian, then the polynomial ring over R in one variable, $R[x]$, is also Noetherian.

The details of the following proof are left as an exercise.

Sketch of Proof: Take any ideal $J \subset R[x]$ and define $I_i \subset R$ to be the ideal consisting of those elements $a_i \in R$ that are leading coefficients of some degree-i polynomial

$$a_i x^i + \cdots + a_1 x + a_0 \in J.$$

The ideals $I_i \subset R$ form an increasing sequence

$$I_0 \subset I_1 \subset \cdots .$$

By the Noetherian property of R (see the exercise at the end of the section), eventually we have equality,

$$I_0 \subset I_1 \subset \cdots \subset I_r = I_{r+1} = \cdots .$$

For $i = 0, \ldots, r$, take generators $a_{i1}, \ldots, a_{in_i}$ for I_i. Choose degree-i polynomials $F_{ij} \in J$ ($i = 0, \ldots, r$ and $j = 1, \ldots, n_i$) with leading coefficient a_{ij}. By induction with respect to the degree of $f \in J$ we can prove that J is generated by the polynomials F_{ij}. $\qquad\square$

Hilbert's Basis Theorem immediately implies that any polynomial ring over a Noetherian ring R is also Noetherian. This follows from induction on the number of variables, using the fact that $R[x_1, \ldots, x_n] = R[x_1, \ldots, x_{n-1}][x_n]$. In particular, the fundamental ring of algebraic geometry, $\mathbb{C}[x_1, \ldots, x_n]$, is Noetherian.[1] We only need to check that $\mathbb{C}$ is Noetherian, and this is obvious because a field has only two ideals, the zero ideal and the unit ideal (which is generated by 1).

Now we turn to an important application. Consider an affine algebraic variety V in $\mathbb{A}^n$. We claim that the set

$$\mathbb{I}(V) = \{f \in \mathbb{C}[x_1, \ldots, x_n] \mid f(x) = 0 \text{ for all } x \in V\}$$

is an ideal of $\mathbb{C}[x_1, \ldots, x_n]$. Indeed, if f and g both vanish on V, then clearly $f + g$ vanishes on V; likewise, if f vanishes on V, and r is an arbitrary polynomial, then rf vanishes on V. Thus $\mathbb{I}(V)$ is an ideal.

Now, by definition, V is contained in $\mathbb{V}(\mathbb{I}(V))$. On the other hand, it is also easy to see that $\mathbb{V}(\mathbb{I}(V))$ is contained in V. Indeed, if $x \in \mathbb{V}(\mathbb{I}(V))$, then $f(x) = 0$ for all $f \in \mathbb{I}(V)$. But because V is defined by the vanishing of $\{F_i\}_{i \in I}$, evidently $F_i \in \mathbb{I}(V)$, so x is in the common zero set of the F_i. Therefore, for all affine varieties V in $\mathbb{A}^n$, we have

$$\mathbb{V}(\mathbb{I}(V)) = V.$$

Now we can put our observations together to conclude that every algebraic variety can be described as the common zero locus of finitely many polynomials. Because $\mathbb{C}[x_1, \ldots, x_n]$ is Noetherian, the ideal $\mathbb{I}(V)$ of polynomials vanishing on $V \subseteq \mathbb{A}^n$ is finitely generated, say

$$\mathbb{I}(V) = (F_1, \ldots, F_r).$$

So by the previous remarks,

$$V = \mathbb{V}(\mathbb{I}(V)) = \mathbb{V}((F_1, \ldots, F_r)) = \mathbb{V}(F_1, \ldots, F_r),$$

and V is the set of common zeros of the finite collection of polynomials $F_1, \ldots, F_r$.

---

[1] In algebraic number theory there is a similarly fundamental Noetherian ring, the ring of integers $\mathbb{Z}$.

The fact that every affine algebraic variety can be described by finitely many polynomials is an important and useful fact.

Historical Remarks: Hilbert was motivated by his interest in *invariant theory*, the study of those polynomials left invariant under the action of some group of linear transformations contained in $\mathbf{GL}(n)$. The basis theorem implies the finite generation of the ring of polynomials invariant under the action of a finite group; see [9, Section 1.4.1]. This had been viewed as the central problem in invariant theory at the time Hilbert's paper appeared in 1890. Hilbert's lecture notes are still a good introduction to invariant theory [23].

Because invariant theory had been primarily concerned with the explicit computation of bases, Hilbert's nonconstructive proof was controversial. Paul Gordan, the leading expert in invariant theory at the time, exclaimed, "This is not mathematics, this is theology!" When Hilbert refined his ideas to produce a method that could (theoretically) be used to compute generators, Gordan was forced to concede, "Theology also has its advantages." See Reid's entertaining biography of Hilbert [35].

The problem of whether or not the ring of invariants is finitely generated for any group G was proposed by Hilbert in his famous speech at the 1900 International Congress of Mathematicians. This problem came to be known as Hilbert's fourteenth problem, and remained open until the late fifties, when Nagata found a ring of invariants that is not finitely generated.

The existential approach taken by Hilbert dealt a powerful blow to computational algebra, as mathematicians quickly turned to more abstract methods. With the advent of the computer, computational methods have recently resumed their position in the mainstream of mathematical research. For a nice introduction to this topic, see [5] and its companion volume [6].

Exercise 2.2.1. Show that a ring R is Noetherian if and only if every strictly ascending sequence of ideals $I_1 \subsetneq I_2 \subsetneq \cdots$ is finite.

Exercise 2.2.2. Show that every affine algebraic variety is the intersection of finitely many hypersurfaces.

Exercise 2.2.3. Let the group S_3 of permutations of three letters act on the polynomial ring $\mathbb{C}[x_1, x_2, x_3]$ by permutation of the variables. Find the ring of invariant polynomials.

2.3   Hilbert's Nullstellensatz

We now turn to a fundamental theorem of algebraic geometry, Hilbert's Nullstellensatz.

We have seen that the set of polynomials vanishing on an affine algebraic variety V forms an ideal in the polynomial ring. However, such ideals are of a special sort: They are radical.[2] Indeed, if f is a polynomial such that f^n vanishes on V, then for all $x \in V$, $f^n(x) = (f(x))^n = 0$. This means that $f(x) = 0$ as well, and f also vanishes on V. This proves that $\mathbb{I}(V)$, the ideal of all polynomials vanishing on V, is a radical ideal.

We have already seen that $V = \mathbb{V}(\mathbb{I}(V))$ for any affine algebraic variety V. Hilbert's Nullstellensatz states that the mappings $V \mapsto \mathbb{I}(V)$ and $I \mapsto \mathbb{V}(I)$ are essentially inverse to each other, at least if we restrict our attention to radical ideals I. This famous theorem is the first entry in a dictionary that will help us translate statements about geometry into the language of algebra.

Hilbert's Nullstellensatz: For any ideal $I \subset \mathbb{C}[x_1, \ldots, x_n]$,

$$\mathbb{I}(\mathbb{V}(I)) = \sqrt{I}.$$

In particular, if I is radical, then

$$\mathbb{I}(\mathbb{V}(I)) = I.$$

Proof: See any book on commutative algebra, for instance [9, page 134, (also see pages 142–144)], or books on algebraic geometry, for example, [17, page 57].   □

Hilbert's Nullstellensatz implies a one–to–one correspondence:

$$\left\{ \begin{array}{c} \text{affine algebraic} \\ \text{varieties in } \mathbb{A}^n \end{array} \right\} \longleftrightarrow \left\{ \begin{array}{c} \text{radical ideals} \\ \text{in } \mathbb{C}[x_1, \ldots, x_n] \end{array} \right\}.$$

Note that if V is a subvariety of W, then functions vanishing on W are forced to vanish on V, so $\mathbb{I}(W) \subset \mathbb{I}(V)$. Thus, Hilbert's correspondence is order-reversing.

The order-reversing correspondence given by Hilbert's Nullstellensatz implies that every maximal ideal in the polynomial ring $\mathbb{C}[x_1, \ldots, x_n]$ is the ideal of functions vanishing at a single point $(a_1, \ldots, a_n) \in \mathbb{A}^n$. In particular, each maximal ideal has the form $\mathfrak{m}_a = (x_1 - a_1, \ldots, x_n - a_n)$, and the corresponding variety is the singleton $\mathbb{V}(\mathfrak{m}_a) = \{a\} = \{(a_1, \ldots, a_n)\} \subset \mathbb{A}^n$. In other words, Hilbert's Nullstellensatz identifies the set of maximal ideals of the polynomial ring $\mathbb{C}[x_1, \ldots, x_n]$ with the points in affine space $\mathbb{A}^n$.

Hilbert's Nullstellensatz can be viewed as a multidimensional version of the Fundamental Theorem of Algebra. The ideal generated by a single polynomial in one variable is radical if and only if it has no repeated roots. The Fundamental Theorem amounts to the fact that a radical ideal in

---

[2]Recall that an ideal I is radical if $r^n \in I$ implies that $r \in I$. See Section 2.1.

$\mathbb{C}[z]$ is completely determined by the zero set of a generator. Hilbert's Nullstellensatz says that a radical ideal $I \subset \mathbb{C}[x_1, \dots, x_n]$ is completely determined by its zero set $\mathbb{V}(I)$.

A natural question that arises regarding Hilbert's Nullstellensatz is whether it can be made "effective." Consider an ideal $I = (F_1, \dots, F_r) \subset \mathbb{C}[x_1, \dots, x_n]$ and the corresponding variety $V = \mathbb{V}(I) \subset \mathbb{A}^n$. By the Nullstellensatz $\sqrt{I} = \mathbb{I}(V)$. So, if $g(x) = 0$ for all $x \in V$, then $g^M \in I$ for some $M > 0$. Can we bound the exponent M in terms of, say, the degrees of the polynomials F_i? What is the smallest possible M that works in general? Until recently, very little was known about such an "effective Nullstellensatz," but in 1988, János Kollár provided a virtually conclusive answer. For example, Kollár shows that if I is generated by r homogeneous[3] polynomials F_i of degree $d_i > 2$, then

$$g \in \sqrt{I} \implies g^M \in I$$

for some $M \le \prod_{i=1}^{r} d_i$. If $r < n$, this result is sharp: No smaller value of M will work in general. Kollár also finds sharp bounds for M when $r \ge n$; see [26].

Algebraic geometry over fields other than $\mathbb{C}$: Algebraic geometry can be applied to the zero loci of polynomials over fields other than $\mathbb{C}$. Given any field $\mathbb{K}$, we can study the zero sets in $\mathbb{K}^n$ of polynomials with coefficients in $\mathbb{K}$. The Zariski topology is defined in $\mathbb{K}^n$. The set of polynomials vanishing on a variety in $\mathbb{K}^n$ forms an ideal in $\mathbb{K}[x_1, \dots, x_n]$, and this ideal is finitely generated. Thus, much of the basic machinery we have discussed goes through unchanged. However, there are some serious difficulties with Hilbert's Nullstellensatz.

Like the Fundamental Theorem of Algebra, Hilbert's Nullstellensatz fails over the real numbers. For instance, it is easy to check that the ideal (x^2+1) is a radical ideal in the ring $\mathbb{R}[x]$, since $\mathbb{R}[x]/(x^2+1) \cong \mathbb{C}$ is a field. The real null set $\mathbb{V}(x^2+1)$ of this ideal is empty, so it coincides with the null set of the trivial ideal generated by 1 in $\mathbb{R}[x]$. Thus, two different radical ideals define the same variety in $\mathbb{R}^n$, and Hilbert's Nullstellensatz fails.

However, Hilbert's Nullstellensatz holds for varieties defined over any algebraically closed field. (Recall that a field $\mathbb{K}$ is said to be algebraically closed if every nonconstant polynomial with coefficients in $\mathbb{K}$ has a root in $\mathbb{K}$.) Provided that $\mathbb{K}$ is algebraically closed, Hilbert's Nullstellensatz guarantees a one-to-one correspondence between subvarieties of $\mathbb{K}^n$ and the radical ideals of the polynomial ring $\mathbb{K}[x_1, \dots, x_n]$. This holds, for example, for fields as exotic as $\overline{\mathbb{F}}_2$, the algebraic closure of the field with two elements.

---

[3]A polynomial is *homogeneous* when its terms all have the same degree; see Section 3.2 for the geometric importance of homogeneous polynomials.

Algebraic geometry over non-algebraically closed fields (and especially $\mathbb{R}$) is a difficult and active area of research today. See, for example, [28] and [38].

Besides the fact that $\mathbb{C}$ is algebraically closed, another property of $\mathbb{C}$ that is often very useful is that $\mathbb{C}$ is a field of *characteristic zero*, that is, the integers $\mathbb{Z}$ form a subring of $\mathbb{C}$. For a prime number p, we say that a field has characteristic p if it contains the field $\mathbb{F}_p$ of p elements as a subring; otherwise, we say that it has characteristic zero. The field $\overline{\mathbb{F}_2}$ above has characteristic two. We will not treat fields of nonzero characteristic in this book, although occasionally we point out where trouble may arise if the field does not contain the integers.

Exercise 2.3.1. Check that prime ideals correspond to irreducible varieties. (Recall that a variety V is irreducible if it cannot be decomposed as the union of two distinct proper subvarieties.) Check that the ideal (xy, xz) defines a reducible variety and is radical but not prime.

Exercise 2.3.2. Show that the dimension of an affine algebraic variety is finite.

Exercise 2.3.3. Show that a radical ideal I in the ring $\mathbb{C}[x_1, \ldots, x_n]$ is the intersection of all the maximal ideals $(x_1 - a_1, \ldots, x_n - a_n)$ containing I.

Exercise 2.3.4. Prove that the Zariski topology on an affine algebraic variety is compact: Every open cover has a finite subcover.

Exercise 2.3.5. Prove that the complement of a point in $\mathbb{A}^n$ is an open set that is compact in the Zariski topology.

Exercise 2.3.6. Show that the zero set in $\mathbb{A}^2$ of the function $y - e^x$ is not an affine algebraic variety.

2.4   The Coordinate Ring

One theme in modern mathematics is that in order to understand certain objects we ought to study natural classes of functions on them. In topology we study continuous functions on topological spaces, in differential geometry we study smooth functions on manifolds, and in complex geometry we study holomorphic functions on complex manifolds. In algebraic geometry, varieties are defined by polynomials, and it is most appropriate to look at polynomial functions on them.

Let $V \subset \mathbb{A}^n$ be an affine algebraic variety. Given any complex polynomial in n variables, the restriction to V defines a function $V \to \mathbb{C}$. Under the usual pointwise operations of addition and multiplication, these functions naturally form a $\mathbb{C}$-algebra

$$\mathbb{C}[x_1, \ldots, x_n]|_V,$$

which we call the *coordinate ring* of V and denote by $\mathbb{C}[V]$. In particular, the coordinate ring of affine space $\mathbb{A}^n$ is the polynomial ring $\mathbb{C}[\mathbb{A}^n] = \mathbb{C}[x_1, \ldots, x_n]$.

The elements of $\mathbb{C}[V]$ are restrictions of polynomials on $\mathbb{A}^n$, but we usually denote them by the original polynomials. This can be slightly confusing, since two different polynomials may well have the same restriction to V. For example, the zero polynomial and the polynomial $P(x, y, z) = x^2 + y^2 + z^2$ obviously restrict to the same function on the variety $\mathbb{V}(x^2 + y^2 + z^2)$ in $\mathbb{A}^3$ defined by the vanishing of P.

Obviously, restriction defines a surjective ring homomorphism

$$\mathbb{C}[x_1, \ldots, x_n] \to \mathbb{C}[x_1, \ldots, x_n]|_V$$

with kernel precisely the ideal of functions $\mathbb{I}(V)$ vanishing on V. So the coordinate ring $\mathbb{C}[V]$ is isomorphic to the ring

$$\frac{\mathbb{C}[x_1, \ldots, x_n]}{\mathbb{I}(V)}$$

in a natural way. Equivalence classes in $\mathbb{C}[V]$ correspond to functions on V. Each equivalence class is usually denoted by some representative, a polynomial like x, y, x_1, or $x^2 + xy$.

Sometimes we write the restriction of a polynomial in a form that hides its polynomial character. As an example, think of the function $\frac{1}{x}$ defined on the variety $V = \mathbb{V}(xy - 1) \subset \mathbb{A}^2$. Because $xy = 1$ everywhere on V, the function $\frac{1}{x}$ is evidently the same as the restriction of the polynomial function y to V.

Example: Consider the cone $V = \mathbb{V}(x^2 + y^2 - z^2)$ in $\mathbb{A}^3$. Because the polynomial $x^2 + y^2 - z^2$ is irreducible, it generates a prime, and hence radical, ideal. By Hilbert's Nullstellensatz, $\mathbb{I}(V)$ is generated by the polynomial $x^2 + y^2 - z^2$. The coordinate ring of our cone is therefore the quotient ring $\mathbb{C}[x, y, z]/(x^2 + y^2 - z^2)$. It is customary to call this ring $\mathbb{C}[x, y, z]$ *equipped with the relation* $x^2 + y^2 - z^2 = 0$, meaning just that the polynomial $x^2 + y^2 - z^2$ can be interpreted as zero, wherever it appears. For example, in this ring we have

$$x^3 + 2xy^2 - 2xz^2 + x = 2x(x^2 + y^2 - z^2) + x - x^3 = x - x^3.$$

Just as each affine algebraic variety determines a unique $\mathbb{C}$-algebra (its coordinate ring), every morphism of affine varieties determines a unique $\mathbb{C}$-algebra homomorphism between the corresponding $\mathbb{C}$-algebras.

Indeed, given any morphism $V \xrightarrow{F} W$ of affine algebraic varieties, there is a naturally induced map of coordinate rings

$$\mathbb{C}[W] \longrightarrow \mathbb{C}[V],$$
$$g \longmapsto g \circ F,$$

called the *pullback* of F, given by composing a function g on W with F. It is easy to check that the pullback $g \circ F$ of a polynomial function g on W is indeed a polynomial function on V because the map $V \xrightarrow{F} W$ is itself given by polynomials, and the composition of two polynomials is again a polynomial. It is also easy to check that this pullback map defines a $\mathbb{C}$-algebra homomorphism from $\mathbb{C}[W]$ to $\mathbb{C}[V]$.

Example: Consider the morphism of algebraic varieties

$$\mathbb{A}^3 \longrightarrow \mathbb{A}^2,$$
$$(x, y, z) \longmapsto (x^2 y, x - z).$$

Letting (u, v) denote the coordinates of $\mathbb{A}^2$, the pullback defines a map

$$\mathbb{C}[u, v] \longrightarrow \mathbb{C}[x, y, z],$$
$$u \longmapsto x^2 y,$$
$$v \longmapsto x - z.$$

Note that this $\mathbb{C}$-algebra map is completely determined by where it sends the generators u and v. For example, the polynomial $u^2 + v^3$ in $\mathbb{C}[u, v]$ is sent to the polynomial $(x^2 y)^2 + (x - z)^3$ in $\mathbb{C}[x, y, z]$.

Example: The pullback is a generalization of the dual map in linear algebra. To make this clear we look at a morphism F with homogeneous linear components $F_1, \ldots, F_m$. The map F is a linear mapping $\mathbb{C}^n \to \mathbb{C}^m$, whose matrix is made up of the coefficients of the linear forms F_i. The pullback operation on the coordinate ring can be restricted to the linear functionals (degree-1 polynomials). Denote the restriction of the pullback of F by F^*. Then F^* defines a vector space map $(\mathbb{C}^m)^* \to (\mathbb{C}^n)^*$, and this is the standard dual map in linear algebra.

Exercise 2.4.1. Prove that the coordinate ring of an affine algebraic variety is a reduced, finitely generated $\mathbb{C}$-algebra. (Recall that a ring is said to be reduced if it has no nonzero nilpotent elements.)

Exercise 2.4.2. Remember that a radical ideal in the ring $\mathbb{C}[x_1, \ldots, x_n]$ is the intersection of all the maximal ideals containing it. Now prove the same statement for a radical ideal in the coordinate ring $\mathbb{C}[V]$ by considering the correspondence between ideals in a quotient ring $\frac{R}{I}$ and ideals of R containing I.

2.5   The Equivalence of Algebra and Geometry

We have seen that each affine algebraic variety V determines a unique $\mathbb{C}$-algebra $\mathbb{C}[V]$, its coordinate ring, and that each morphism $V \to W$ of affine algebraic varieties determines a unique $\mathbb{C}$-algebra homomorphism $\mathbb{C}[W] \to \mathbb{C}[V]$, its pullback. The defining feature of algebraic geometry is the remarkable fact that not only does the geometry determine the algebra, but conversely, the algebra determines the geometry. That is, given any finitely generated $\mathbb{C}$-algebra R without nilpotent elements, there exists an affine algebraic variety V, uniquely defined up to isomorphism, such that R is isomorphic to the coordinate ring of V. Moreover, any homomorphism between such $\mathbb{C}$-algebras uniquely defines a morphism of the corresponding varieties. In fancy language, there is an *equivalence of categories* between the category of affine algebraic varieties and finitely generated, reduced $\mathbb{C}$-algebras. Our next task is to explain this equivalence.

To start, note that the coordinate ring $\mathbb{C}[V] = \frac{\mathbb{C}[x_1,\ldots,x_n]}{\mathbb{I}(V)}$ of an affine algebraic variety is a finitely generated reduced $\mathbb{C}$-algebra. The functions $x_1,\ldots,x_n$ are $\mathbb{C}$-algebra generators for $\mathbb{C}[V]$, and since $\mathbb{I}(V)$ is a radical ideal, the quotient ring $\mathbb{C}[V]$ has no nilpotent elements (that is, $\mathbb{C}[V]$ is reduced).

Conversely, every reduced finitely generated $\mathbb{C}$-algebra R is isomorphic to the coordinate ring of some variety. To see this, fix any finite set of $\mathbb{C}$-algebra generators for R and note that R is isomorphic to $\mathbb{C}[x_1,\ldots,x_n]/I$, where I is the kernel of the surjective homomorphism

$$\mathbb{C}[x_1,\ldots,x_n] \;\longrightarrow\; R,$$
$$x_j \;\longmapsto\; j\text{th generator of } R.$$

Because the quotient ring $\mathbb{C}[x_1,\ldots,x_n]/I$ is reduced, the ideal I is a radical ideal of $\mathbb{C}[x_1,\ldots,x_n]$. Hence the ideal I defines a variety $\mathbb{V}(I) \subseteq \mathbb{A}^n$ whose coordinate ring is isomorphic to R.

Furthermore, we have seen that a morphism of algebraic varieties $V \to W$ gives rise to a $\mathbb{C}$-algebra homomorphism $\mathbb{C}[W] \to \mathbb{C}[V]$ by the pullback. Conversely, every $\mathbb{C}$-algebra homomorphism between finitely generated reduced $\mathbb{C}$-algebras is the pullback of a uniquely defined morphism between the corresponding varieties, as we prove below.

Theorem: Every finitely generated reduced $\mathbb{C}$-algebra is isomorphic to the coordinate ring of some affine algebraic variety.

If $V \xrightarrow{F} W$ is a morphism of affine algebraic varieties, then its pullback is a homomorphism between the coordinate rings $\mathbb{C}[W] \xrightarrow{F^\#} \mathbb{C}[V]$.

If

$$R \xrightarrow{\sigma} S$$

is a homomorphism of reduced finitely generated $\mathbb{C}$-algebras, then there exists a morphism F of affine algebraic varieties corresponding to R and S such that σ is the pullback of F. This morphism F is unique up to isomorphism.

Proof: The first part of the theorem just repeats what we have already discussed. What remains to be seen is how we can construct a morphism of affine algebraic varieties from an abstract $\mathbb{C}$-algebra homomorphism.

Fix a $\mathbb{C}$-algebra map $R \xrightarrow{\sigma} S$. Because R and S are finitely generated reduced $\mathbb{C}$-algebras, we may choose presentations for them and write

$$\frac{\mathbb{C}[x_1,\ldots,x_n]}{I} \xrightarrow{\sigma} \frac{\mathbb{C}[y_1,\ldots,y_m]}{J},$$

where I and J are radical ideals. We are looking for a morphism, that is, a polynomial mapping $\mathbb{A}^m \xrightarrow{F} \mathbb{A}^n$ such that F sends the subvariety $V = \mathbb{V}(J)$ of $\mathbb{A}^m$ into the variety $W = \mathbb{V}(I)$. Furthermore, we must have $F^{\#} = \sigma$.

For $j = 1,\ldots,n$, let $F_j \in \mathbb{C}[y_1,\ldots,y_m]$ be any polynomials representing the image $\sigma(x_j)$ of x_j under the map $\mathbb{C}[x_1,\ldots,x_n]/I \xrightarrow{\sigma} \mathbb{C}[y_1,\ldots,y_m]/J$. Define the polynomial map

$$\mathbb{A}^m \xrightarrow{F} \mathbb{A}^n,$$
$$a = (a_1,\ldots,a_m) \longmapsto (F_1(a),\ldots,F_n(a)).$$

We claim that F maps V to W. To see this, let $a \in V = \mathbb{V}(J)$. We want to show that $F(a) \in W$. For this, it is sufficient to check that $F(a)$ is in the zero set of every polynomial G in I. Using the fact that $F_j = \sigma(x_j)$, we see that for G in I,

$$\begin{aligned} G(F(a)) &= G(F_1(a),\ldots,F_n(a)) \\ &= G(\sigma(x_1)(a),\ldots,\sigma(x_n)(a)) \\ &= \sigma(G)(a). \end{aligned}$$

Because $G \in I$, it represents the zero class of $\mathbb{C}[x_1,\ldots,x_n]/I$; so its image $\sigma(G)$ under the ring homomorphism σ must represent the zero class of $\mathbb{C}[y_1,\ldots,y_m]/J$. In other words, $\sigma(G)$ lies in J, the ideal of all functions vanishing on V. Now for $a \in V$, we see that $\sigma(G)(a) = 0$ for all $G \in I$, so that $G(F(a)) = 0$ for all $G \in I$. Thus $F(a) \in W = \mathbb{V}(I)$ and F maps V to W. The reader should verify without difficulty that $F^{\#} = \sigma$.

There was an arbitrariness in the choice of representative F_j of $\sigma(x_j)$; F_j can be replaced by any polynomial F_j' in $\mathbb{C}[y_1,\ldots,y_m]$ that represents the same equivalence class in $\frac{\mathbb{C}[y_1,\ldots,y_m]}{J}$. Since the difference $F_j - F_j'$ vanishes on V, the resulting morphisms of $\mathbb{A}^m$ to $\mathbb{A}^n$ restrict to the same morphism on V.

Finally, we address the point that this morphism is unique up to isomorphism. As constructed, our morphism appears to be unique, and in fact, it *is* unique once we fix the presentation of our original algebras in the form

$$\frac{\mathbb{C}[x_1, \ldots, x_n]}{I}.$$

A different set of algebra generators for the algebras R and S would produce different ideals of relations (in possibly a different number of variables). Therefore, the resulting varieties V' and W' would be different from (but isomorphic to) V and W, and so would the corresponding morphism $V' \xrightarrow{F'} W'$. However, F is isomorphic to F' in the sense that the following diagram commutes:

$$
\begin{array}{ccc}
V & \xrightarrow{\ F\ } & W \\
\cong \downarrow & & \downarrow \cong \\
V' & \xrightarrow[F']{} & W'
\end{array}
$$

We leave it to the reader to chase through the definitions and verify this.   □

The theorem shows that affine varieties and their morphisms are essentially equivalent to finitely generated reduced $\mathbb{C}$-algebras and their homomorphisms, only with the arrows reversed. In other words, the *categories* of affine algebraic varieties and of finitely generated reduced $\mathbb{C}$-algebras are equivalent (or *anti-isomorphic* if we want to emphasize the order-reversing feature of the equivalence). Obviously, $(F \circ G)^{\#} = G^{\#} \circ F^{\#}$ whenever defined, so forming the pullback is in itself a kind of direction-reversing "homomorphism," a so-called *contravariant functor*. This is essential for the identification of categories.

As a consequence of the theorem, we see that two varieties are isomorphic if and only if their coordinate rings are isomorphic. The next few examples illustrate the usefulness of this idea.

Example: We have seen that the morphism

$$
\begin{array}{ccc}
\mathbb{A}^1 & \longrightarrow & \mathbb{V}(y - x^2) \subset \mathbb{A}^2, \\
t & \longmapsto & (t, t^2),
\end{array}
$$

is an isomorphism. See Figure 2.1.

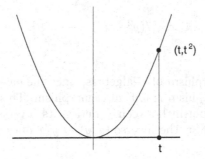

Figure 2.1. The parabola and the line are isomorphic

Notice that the pullback

$$\mathbb{C}[x,y]/(y - x^2) \longrightarrow \mathbb{C}[t],$$
$$x \longmapsto t,$$
$$y \longmapsto t^2,$$

is surjective with zero kernel, and hence it is an algebra isomorphism. Alternatively, one can check that projection onto the first coordinate $(t, t^2) \mapsto t$ defines an inverse morphism.

This example should be compared with

$$\mathbb{A}^1 \longrightarrow \mathbb{V}(y^2 - x^3) \subset \mathbb{A}^2,$$
$$t \longmapsto (t^2, t^3),$$

which is bijective but not an isomorphism. Here we should think of $t \in \mathbb{A}^1$ as the slope of a line $L(t)$ passing through the origin; the line $L(t)$ meets the curve $\mathbb{V}(y^2 - x^3)$ in another point, (t^2, t^3). See Figure 2.2.

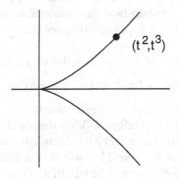

Figure 2.2. $\mathbb{V}(y^2 - x^3)$

Here the pullback is

$$
\begin{aligned}
\mathbb{C}[x,y]/(y^2 - x^3) &\longrightarrow \mathbb{C}[t], \\
x &\longmapsto t^2, \\
y &\longmapsto t^3,
\end{aligned}
$$

clearly not an isomorphism of $\mathbb{C}$-algebras, since the element t is not in the image. Thus our morphism is not an isomorphism. This seems reasonable: If it were an isomorphism between varieties, its inverse would be an isomorphism too. However, the inverse $(x,y) \mapsto y/x$ does not appear to be a polynomial.

Exercise 2.5.1. Show that the pullback $\mathbb{C}[W] \xrightarrow{F^{\#}} \mathbb{C}[V]$ is injective if and only if F is *dominant*; that is, the image set $F(V)$ is dense in W.

Exercise 2.5.2. Show that the pullback $\mathbb{C}[W] \xrightarrow{F^{\#}} \mathbb{C}[V]$ is surjective if and only if F defines an isomorphism between V and some algebraic subvariety of W.

Exercise 2.5.3. If $F = (F_1, \ldots, F_n) : \mathbb{A}^n \to \mathbb{A}^n$ is an isomorphism, then show that the *Jacobian determinant*

$$
\det \begin{bmatrix}
\frac{\partial F_1}{\partial x_1} & \cdots & \frac{\partial F_1}{\partial x_n} \\
\vdots & & \vdots \\
\frac{\partial F_n}{\partial x_1} & \cdots & \frac{\partial F_n}{\partial x_n}
\end{bmatrix}
$$

is a nonzero constant polynomial. It is not known whether the converse is true. This is a famous open problem known as the *Jacobian conjecture*.

2.6   The Spectrum of a Ring

As we have seen, Hilbert's Nullstellensatz allows us to identify the points of any affine algebraic variety V with the maximal ideals of its coordinate ring $\mathbb{C}[V]$. We now want to explain how the maximal ideals of any commutative ring can be considered as a topological space that is like a variety in many respects.

The *maximal spectrum* of a ring R is the set of maximal ideals in R:

$$
\mathrm{maxSpec}\, R = \{ \mathfrak{m} \subset R \mid \mathfrak{m} \text{ is a maximal ideal} \}.
$$

The identification of an affine algebraic variety V with the maximal spectrum of its coordinate ring $\mathrm{maxSpec}\,\mathbb{C}[V]$ is deeper than a mere set-theoretic correspondence. We can transport the Zariski topology on V to a topology on $\mathrm{maxSpec}\,\mathbb{C}[V]$ as follows: The points of a Zariski-closed set $W \subset V$ correspond to the set of maximal ideals in the coordinate ring $\mathbb{C}[V]$ that contain the ideal $\mathbb{I}(W)$ corresponding to W. In other words, the closed sets of the Zariski topology on $\mathrm{maxSpec}\,\mathbb{C}[V]$ are sets of maximal ideals of

$\mathbb{C}[V]$ containing some given ideal of $\mathbb{C}[V]$. This approach defines the Zariski topology on $\mathrm{maxSpec}\mathbb{C}[V]$ without any direct reference to varieties.

Similarly, consider a morphism of algebraic varieties $V \xrightarrow{F} W$, interpreted as a map $\mathrm{maxSpec}\mathbb{C}[V] \xrightarrow{F} \mathrm{maxSpec}\mathbb{C}[W]$. Hilbert's Nullstellensatz allows us to reconstruct the mapping F from its pullback $\mathbb{C}[W] \xrightarrow{F^\#} \mathbb{C}[V]$. Indeed, given a point p of V, we think of it as a maximal ideal $\mathfrak{m}$ in $\mathrm{maxSpec}\mathbb{C}[V]$. Then the image of p under F corresponds to the maximal ideal $(F^\#)^{-1}(\mathfrak{m})$ in $\mathrm{maxSpec}\mathbb{C}[W]$, as the reader should check. Thus, any homomorphism $R \xrightarrow{\sigma} S$ of finitely generated reduced $\mathbb{C}$-algebras induces a map of associated spectra

$$\mathrm{maxSpec}(S) \longrightarrow \mathrm{maxSpec}(R),$$
$$\mathfrak{m} \longmapsto \sigma^{-1}(\mathfrak{m}).$$

Our success in identifying an algebraic variety with the set of maximal ideals in a suitable ring encourages us to try to develop a theory of algebraic geometry on the set of maximal ideals in any ring.

Given any commutative ring R, we can equip its maximal spectrum $\mathrm{maxSpec}R$ with the Zariski topology by defining the closed sets to be the sets

$$\mathbb{V}(I) = \{\mathfrak{m} \in \mathrm{maxSpec}R \mid \mathfrak{m} \supset I\},$$

where I is an ideal in R. This gives rise to a topological space, but unfortunately it is not exactly what we want. We would like our generalization to mimic the case above. For instance, if $R \xrightarrow{\sigma} S$ is a homomorphism of rings, we would like

$$\mathrm{maxSpec}S \longrightarrow \mathrm{maxSpec}R,$$
$$\mathfrak{m} \longmapsto \sigma^{-1}(\mathfrak{m}),$$

to be a well-defined, continuous map of topological spaces. But unfortunately, the inverse image of a maximal ideal under an arbitrary ring homomorphism need not be a maximal ideal. An example is given by the inclusion $\mathbb{Z} \hookrightarrow \mathbb{Q}$. The inverse image of the maximal ideal $\{0\}$ in $\mathbb{Q}$ is the prime ideal $\{0\}$ in $\mathbb{Z}$, which is not maximal.

However, the inverse image of any prime ideal under an arbitrary ring homomorphism is prime, an easy fact that we leave as an exercise. This suggests that instead of focusing on the set of maximal ideals of R, we ought to switch our attention to the larger set of all prime ideals.

Definition: The *spectrum* $\mathrm{Spec}R$ of a commutative ring R is the set of all of its prime ideals. We equip the spectrum $\mathrm{Spec}R$ with a (Zariski) topology by declaring the closed sets to be the sets of the form $\mathbb{V}(I) = \{\mathfrak{p} \in \mathrm{Spec}R \mid \mathfrak{p} \supset I\}$, where I is an ideal of R. This turns the spectrum into a topological space that contains the maximal spectrum as a topological subspace.

The spectrum of a ring, equipped with its Zariski topology, is what Grothendieck called an *affine scheme*. [4] The theory of schemes revolutionized algebraic geometry, and Grothendieck was awarded a Fields medal in 1966 for this huge body of work. Serious students of algebraic geometry must eventually struggle with the massive tome fondly known as "EGA", *Eléments de Géométrie Algébrique*, where the theory of schemes is developed.

One of the first and most natural classes of schemes is obtained by considering the maximal ideal space of a finitely generated $\mathbb{C}$-algebra, but without assuming that it is reduced. Even if one is primarily interested in varieties, one is often led to consider at least these very special types of schemes. In this book we will occasionally mention these types of schemes for cultural purposes, although they will not be central to our discussion.

The idea of a scheme can be used in algebraic number theory to deal with rings like $R = \frac{\mathbb{Z}[x,y,z]}{(x^n+y^n-z^n)}$. Considering $\mathrm{Spec}R$ leads one to the study of *arithmetic geometry* and ultimately to Wiles's celebrated proof of Fermat's Last Theorem. It is a great tribute to the unity of mathematics that the subject of algebraic geometry, fundamental also in the study of Riemann surfaces, is applicable to arithmetic questions.

Exercise 2.6.1. Prove that the spectrum $\mathrm{Spec}R$ of a commutative ring can be given the structure of a topological space whose closed sets are of the form $\mathbb{V}(I) = \{P \in \mathrm{Spec}R | P \supset I\}$, for I an ideal in R.

Exercise 2.6.2. Prove that a point in $\mathrm{Spec}R$ is closed if and only if it is a maximal ideal.

Exercise 2.6.3. Prove that the maximal spectrum of the ring of integers $\mathbb{Z}$ consists exactly of the ideals generated by prime numbers: $\mathrm{maxSpec}\mathbb{Z} = \{(2), (3), (5), (7), \ldots\}$. Prove that the only other prime ideal $(0) = \{0\}$ is a *dense point* in $\mathrm{Spec}\mathbb{Z}$, that is, it is contained in every nonempty open set. (The singleton $\{(0)\}$ is, of course, compact, but you have now proved that rather than being closed, it is in fact dense.)

Exercise 2.6.4. Let R be the quotient ring $\frac{\mathbb{C}[x,y]}{(x^2)}$. Prove that the topological space $\mathrm{maxSpec}R$ is homeomorphic to $\mathbb{A}^1$. This example illustrates the flavor of a scheme: We ought to think of $\mathrm{maxSpec}R$ as the y-axis $\mathbb{V}(x^2) \subset \mathbb{A}^2$ "counted twice," since it is defined by x^2 instead of the radical ideal generated by x.

Exercise 2.6.5. Fix a complex number $t \in \mathbb{C}$. Describe the scheme $\mathrm{Spec}\frac{\mathbb{C}[x,y]}{(x(x-t))}$. How does it vary with t? What happens as t approaches zero?

---

[4]Strictly speaking, an affine scheme comes equipped with a "sheaf of rings" (see the appendix), but because this additional data is completely determined by the ring, we are not abusing terminology very much by ignoring it.

3
Projective Varieties

3.1 Projective Space

Affine space $\mathbb{A}^n$ has a natural compactification, the projective space $\mathbb{P}^n$, obtained by adding an infinitely distant point in every direction. The goal of this chapter is to introduce projective space and projective varieties and to interpret them as natural compactifications of affine varieties.

Definition: The *Projective n-space,* denoted by $\mathbb{P}^n$, is the set of all one-dimensional subspaces of the vector space $\mathbb{C}^{n+1}$. That is, $\mathbb{P}^n$ is the set of all complex lines through the origin in $\mathbb{C}^{n+1}$.

Of course, projective n-space can be defined over any field $\mathbb{K}$ as the set of one-dimensional subspaces of the vector space $\mathbb{K}^{n+1}$. While projective spaces over fields other than $\mathbb{C}$ are important in algebraic geometry, even if one is primarily interested in complex algebraic varieties, for the sake of concreteness we focus our attention on the case where the ground field is the field of complex numbers.

Projective n-space can be interpreted as the quotient

$$\mathbb{P}^n = \frac{\mathbb{C}^{n+1} \setminus \{0\}}{\sim},$$

where $\sim$ denotes the equivalence relation of points lying on the same line through the origin: $(x_0, \ldots, x_n) \sim (y_0, \ldots, y_n)$ if and only if there exists a nonzero complex number λ such that $(y_0, \ldots, y_n) = (\lambda x_0, \ldots, \lambda x_n)$.

A point in projective space $\mathbb{P}^n$ can be thought of as an equivalence class

$$[(x_0, \ldots, x_n)] = \{(\lambda x_0, \ldots, \lambda x_n) | \lambda \in \mathbb{C}\},$$

where in this notation at least one of the coordinates $x_0, \ldots, x_n$ must be nonzero. As with any equivalence class, a point $p \in \mathbb{P}^n$ is usually denoted by one of its representatives. To distinguish the class from its representative we write colons between the coordinates of the representing point and call them *homogeneous coordinates* of the point in projective space. We also write the class with square brackets rather than round brackets,

$$[x_0 : x_1 : \cdots : x_n] \in \mathbb{P}^n.$$

This notation emphasizes that the homogeneous coordinates are defined only up to nonzero scalar multiple.

We can think of projective space as the usual complex n-dimensional affine space together with an "infinitely distant point in every direction". This is illustrated by the following examples.

Example: One-dimensional projective space $\mathbb{P}^1$ consists of all complex lines through the origin in $\mathbb{C}^2$. By fixing a reference line—a complex line not through the origin—we can choose a representative for each point p in $\mathbb{P}^1$, namely, the unique point where the reference line meets the line through the origin defining p. Only one point in $\mathbb{P}^1$ fails to have such a representative, namely, the point in projective space corresponding to the unique line through the origin parallel to our reference line. It is natural to call this leftover point in projective space the *point at infinity*. This identifies $\mathbb{P}^1$ with the Riemann sphere:

$$\mathbb{P}^1 = \mathbb{C} \cup \{\infty\},$$
$$[x_0 : x_1] \longmapsto \begin{cases} \frac{x_1}{x_0}, & \text{for } x_0 \neq 0, \\ \infty, & \text{for } x_0 = 0. \end{cases}$$

In Figure 3.1, two different points of $\mathbb{P}^1$ are indicated by the dotted lines; these points are also represented by the two marked intersection points on our fixed bold line labeled $\mathbb{C}^1$. The line labeled "∞" is also a point in $\mathbb{P}^1$, which we think of as "the point at infinity" on the complex line $\mathbb{C}^1$.

We can do a similar thing for the projective plane. See Figure 3.2. Again, by fixing any reference plane not passing through the origin, a typical point in $\mathbb{P}^2$ will have a unique representative on the reference plane. The exceptions consist of lines in $\mathbb{C}^3$ through the origin and lying in the plane parallel to our fixed reference plane. These *points at infinity* make up another copy of $\mathbb{P}^1$. So we can think of $\mathbb{P}^2$ as an ordinary complex plane together with a copy of $\mathbb{P}^1$ at infinity. That is,

$$\mathbb{P}^2 = \mathbb{C}^2 \cup \mathbb{P}^1 = \mathbb{C}^2 \cup \mathbb{C} \cup \{\infty\}.$$

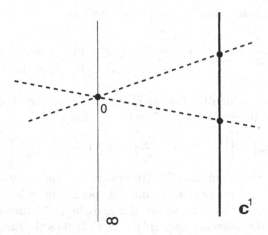

Figure 3.1. The projective line $\mathbb{P}^1$

For example, if we choose coordinates x_0, x_1, x_2 for $\mathbb{C}^3$, so that our reference plane is given by $x_0 = 1$, this identification takes the point $[x_0 : x_1 : x_2]$ to $(\frac{x_1}{x_0}, \frac{x_2}{x_0})$ in the complex plane whenever $x_0 \neq 0$, and to $[x_1 : x_2]$ in the projective line when $x_0 = 0$.

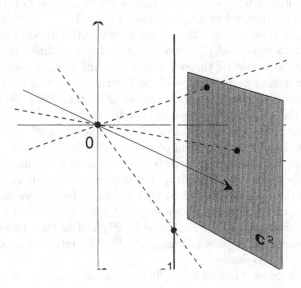

Figure 3.2. The projective plane $\mathbb{P}^2$

Generalizing this idea to arbitrary dimension, we have

$$\mathbb{P}^n \;=\; \mathbb{C}^n \cup \mathbb{P}^{n-1},$$

$$[x_0 : x_1 : \ldots : x_n] \;\longmapsto\; \begin{cases} (\frac{x_1}{x_0}, \ldots, \frac{x_n}{x_0}), & \text{for } x_0 \neq 0, \\ [x_1 : \ldots : x_n], & \text{for } x_0 = 0. \end{cases}$$

With this mapping, we identify the set U_0 in $\mathbb{P}^n$ where the coordinate x_0 is nonzero with the hyperplane $x_0 = 1$ in $\mathbb{C}^{n+1}$, that is,

$$[x_0 : x_1 : \cdots : x_n] = \left[1 : \frac{x_1}{x_0} : \cdots : \frac{x_n}{x_0} \right] \longmapsto \left(1, \frac{x_1}{x_0}, \ldots, \frac{x_n}{x_0} \right),$$

which of course can be identified with $\mathbb{C}^n$. We can think of this copy of $\mathbb{C}^n$ as the "finite" part of $\mathbb{P}^n$. In this case, the remaining points, in which $x_0 = 0$, are called the *points at infinity;* these are the lines in $\mathbb{C}^{n+1}$ through the origin and parallel to the reference hyperplane $x_0 = 1$; thus they naturally form an $(n-1)$-dimensional projective space $\mathbb{P}^{n-1}$.

The choice of x_0 above was arbitrary: We could have done the same with any of the homogeneous coordinates x_i, or indeed with any linear combination of the x_i. In other words, what is "finite" and what is "infinite" is just a matter of perspective. In fact, by defining U_i to be the subset of $\mathbb{P}^n$ where the coordinate x_i is nonzero, we get a useful cover of $\mathbb{P}^n$ by $n+1$ copies of $\mathbb{C}^n$. That is,

$$\mathbb{P}^n = \bigcup_{j=0}^{n} U_j,$$

where $U_j = \{[x_0 : \ldots : x_n] \in \mathbb{P}^n \mid x_j \neq 0\} = \{[x_0 : \ldots : x_n] \in \mathbb{P}^n \mid x_j = 1\}$ can be identified with $\mathbb{C}^n$ (or we may wish to use the notation $\mathbb{A}^n$, since this complex n-space has no preordained origin).

In fact, we need not use the complements of the hyperplanes defined by x_i to define a cover $\{U_i\}$. Given any set of $n+1$ "linearly independent" hyperplanes not passing through the origin, the lines through the origin in $\mathbb{C}^{n+1}$ intersecting the ith hyperplane form a set U_i that can be identified with $\mathbb{C}^n$, and together these give a cover of $\mathbb{P}^n$ that differs from the one we first described only by a change of coordinates in $\mathbb{C}^{n+1}$.

Note that there is a natural Euclidean topology on $\mathbb{P}^n$ induced by virtue of the fact that $\mathbb{P}^n$ is a quotient of $\mathbb{C}^{n+1} - \{0\}$. In particular, two points of $\mathbb{P}^n$ are close together if the corresponding lines in $\mathbb{C}^{n+1}$ have a small angle between them. In this Euclidean topology on $\mathbb{P}^n$, each of the sets U_i is open, and the identification of U_i with $\mathbb{C}^n$ described above defines a homeomorphism of topological spaces when $\mathbb{C}^n$ is considered with its Euclidean topology. Note that each U_i is dense in $\mathbb{P}^n$, and in fact, its complement is a lower-dimensional space (namely, $\mathbb{P}^{n-1}$). The intersection of U_i with U_j when $i \neq j$ is also dense.

The open cover $\{U_i\}$ of $\mathbb{P}^n$ defines an atlas making projective space a complex n-dimensional manifold. The coordinate mappings $U_j \xrightarrow{\psi_j} \mathbb{C}^n$ are essentially the normalized homogeneous coordinates

$$\psi_0: \quad [x_0 : \ldots : x_n] \longmapsto \left(\frac{x_1}{x_0}, \ldots, \frac{x_n}{x_0} \right),$$

$$\psi_1: \quad [x_0 : \ldots : x_n] \longmapsto \left(\frac{x_0}{x_1}, \frac{x_2}{x_1}, \ldots, \frac{x_n}{x_1} \right),$$

$$\vdots$$

$$\psi_n: \quad [x_0 : \ldots : x_n] \longmapsto \left(\frac{x_0}{x_n}, \ldots, \frac{x_{n-1}}{x_n} \right).$$

To convince ourselves that these coordinates make projective space a complex manifold, we have only to check that changes of coordinates $\psi_i(U_i \cap U_j) \overset{\psi_j \circ \psi_i^{-1}}{\longrightarrow} \psi_j(U_i \cap U_j)$ are holomorphic mappings. In fact, they are more than that: They are rational. For instance,

$$\psi_n \circ \psi_0^{-1}(a_1, \ldots, a_n) = \psi_n([1 : a_1 : \ldots : a_n]) = \left(\frac{1}{a_n}, \frac{a_1}{a_n}, \ldots, \frac{a_{n-1}}{a_n} \right).$$

The space $\mathbb{P}^n$ can thus be realized as a complex manifold obtained by gluing together $n + 1$ copies of complex n-space. In fact, because $\mathbb{C}^n$ can be considered instead with its Zariski topology, and the "gluing maps" are rational, it is possible to put the structure of an "abstract algebraic variety" on projective space in a similar way. In the appendix, Section A.1, we define precisely the concept of an abstract algebraic variety, an object very much like a manifold. An abstract algebraic variety is essentially a topological space that has an open cover by affine algebraic varieties, glued together by morphisms of affine algebraic varieties. Rather than embark on this discussion here, we instead show in the next section how the Zariski topology can be defined on projective space in a very concrete way.

Exercise 3.1.1. Show that the complex manifold $\mathbb{P}^n$ is compact.

3.2 Projective Varieties

Before giving any definitions, we remind the reader that there are no non-constant analytic functions on the Riemann sphere $\mathbb{P}^1$. In particular, we cannot expect to find nontrivial polynomial functions on $\mathbb{P}^1$ or on higher-dimensional projective spaces. So we cannot hope to define a projective variety as the common zero set of a collection of polynomial functions on $\mathbb{P}^n$.

We can get around this problem by instead looking at certain kinds of polynomial functions on $\mathbb{C}^{n+1}$. A polynomial $F \in \mathbb{C}[x_0, \ldots, x_n]$ is called *homogeneous* if all its terms have the same degree. The zero set of a homogeneous polynomial in projective space is well-defined. To see this, note that if $F \in \mathbb{C}[x_0, \ldots, x_n]$ is homogeneous of degree d, then

$$F(\lambda x_0, \ldots, \lambda x_n) = \lambda^d F(x_0, \ldots, x_n).$$

Now, if a point $(x_0, \ldots, x_n) \in \mathbb{C}^{n+1}$ is in the zero set of F, then every point of the form $(\lambda x_0, \ldots, \lambda x_n)$, where λ is any constant in $\mathbb{C}$, also lies in the zero set of F. Thus the set of zeros in $\mathbb{C}^{n+1}$ of a homogeneous polynomial is the union of complex lines through the origin. Therefore, although a homogeneous polynomial in $n+1$ variables does not define a function on $\mathbb{P}^n$, it makes sense to speak of its zero set in $\mathbb{P}^n$.

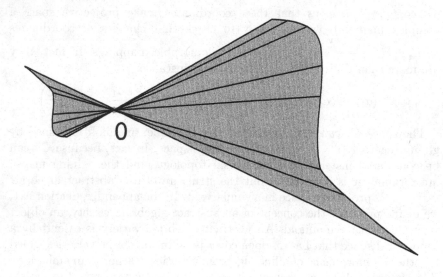

Figure 3.3. Zero set of a homogeneous polynomial

Definition: A projective algebraic variety in $\mathbb{P}^n$ is the common zero set of an arbitrary collection of homogeneous polynomials in $n+1$ variables: $V = \mathbb{V}(\{F_i\}_{i \in I}) \subset \mathbb{P}^n$.

Example: The projective variety $V = \mathbb{V}(x^2 + y^2 - z^2) \subset \mathbb{P}^2$ is called a conic curve.

The conic is the union of its coordinate charts:

$$V = (V \cap U_x) \cup (V \cap U_y) \cup (V \cap U_z).$$

On the chart U_z defined by $z \neq 0$, the conic looks like a complex circle: Identifying U_z with $\mathbb{C}^2$, the curve in U_z is defined by the vanishing of $x^2 + y^2 - 1$. On the charts where x or y is not zero, the same variety is defined by equations that look more like that of a hyperbola, namely

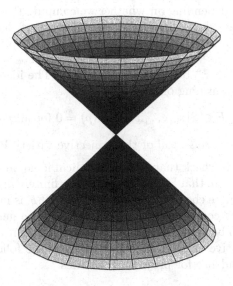

Figure 3.4. $\mathbb{V}(x^2 + y^2 - z^2) \subset \mathbb{P}^2$

$1 + y^2 - z^2 = 0$ and $x^2 + 1 - z^2 = 0$, respectively.[1] In the exercises we will see that any complex conic section can be constructed as an affine chart of this variety.

As in the example, the intersection of any projective variety V with one of the affine coordinate charts of $\mathbb{P}^n$ is an affine algebraic variety. For example, let U_i be the open set of $\mathbb{P}^n$ where the coordinate x_i is not zero, and recall that U_i can be identified with affine space $\mathbb{A}^n$. Then setting the variable x_i to 1 in the defining polynomials for V gives a set of defining polynomials for $V \cap U_i$. Thus, as in the case of projective space itself, we can think of a projective variety as being covered by affine charts:

$$V = (V \cap U_0) \cup (V \cap U_1) \cup \ldots \cup (V \cap U_n),$$
$$V \cap U_i \subset U_i \cong \mathbb{A}^n.$$

Another way to think about a projective variety in $\mathbb{P}^n$ is to imagine a cone-shaped variety in $\mathbb{C}^{n+1}$, but then to identify all points lying on the same line through the origin. The variety in $\mathbb{C}^{n+1}$ defined by a collection of homogeneous polynomials in $x_0, \ldots, x_n$ is called the *affine cone* over the projective variety in $\mathbb{P}^n$ defined by the same homogeneous polynomials. Care is required in writing $\mathbb{V}(F_1, \ldots, F_r)$ to denote a variety defined

---

[1] We often use the words "hyperbola" or "parabola" to describe varieties, but really these words describe only the set of real points on them.

by homogeneous polynomials in $\mathbb{C}[x_0, \ldots, x_n]$, since this notation has two different meanings depending on whether we compute the zero set in $\mathbb{C}^{n+1}$ or in $\mathbb{P}^n$. To distinguish the two cases, we will speak of the "affine variety" or the "projective variety" defined by the F_i's.

Definition: Let $V \subset \mathbb{P}^n$ be a projective variety. The ideal of polynomials in $n + 1$ variables vanishing on V,

$$\mathbb{I}(V) = \{F \in \mathbb{C}[x_0, \ldots, x_n] \mid F(p) = 0 \text{ for all } p \in V\},$$

is called the *homogeneous ideal* of the projective variety V.

It is elementary to check that $\mathbb{I}(V)$ is a radical ideal, and it follows from Hilbert's basis theorem that $\mathbb{I}(V)$ is generated by finitely many polynomials. Furthermore, it can be checked that if a polynomial F is in $\mathbb{I}(V)$, then each of its homogeneous components F_i is in $\mathbb{I}(V)$, so the generators of $\mathbb{I}(V)$ may be assumed to be homogeneous polynomials. By considering the affine cone over a projective variety, it is easy to prove the following version of Hilbert's correspondence for projective varieties.

The Homogeneous Nullstellensatz: The projective subvarieties of $\mathbb{P}^n$ stand in one-to-one correspondence with the radical ideals of the ring $\mathbb{C}[x_0, \ldots, x_n]$ that admit a set of homogeneous generators, with the exception of $(x_0, \ldots, x_n)$ (which defines the origin in the space $\mathbb{C}^{n+1}$).

This motivates us to define the *homogeneous coordinate ring* of a projective variety $V \subset \mathbb{P}^n$ to be the ring

$$\frac{\mathbb{C}[x_0, \ldots, x_n]}{\mathbb{I}(V)}.$$

The homogeneous coordinate ring of a projective variety $V \subseteq \mathbb{P}^n$ is identical to the coordinate ring of the affine cone over V in $\mathbb{A}^{n+1}$. However, we cannot consider the elements of this ring as functions on V.

Just as in the affine case, finite unions of projective varieties in $\mathbb{P}^n$ are projective varieties and arbitrary intersections of projective varieties in $\mathbb{P}^n$ are projective varieties. Thus, the projective varieties in $\mathbb{P}^n$ form the closed sets of a topology, called the *Zariski topology* on $\mathbb{P}^n$. Again, any projective variety in $\mathbb{P}^n$ can be equipped with the subspace topology. The closed sets of this *Zariski topology* on a projective variety are its projective subvarieties.

If V is a projective variety, then there is an induced subspace topology on each of the affine sets $V \cap U_i$ making up V. Fortunately, this induced topology agrees with the Zariski topology on the affine variety $V \cap U_i$, as the reader will check using Exercise 3.2.2 below.

Exercise 3.2.1. Show that every projective variety in $\mathbb{P}^n$ is compact in the induced Euclidean topology. Show that projective varieties are compactifications of affine varieties, in both the Zariski topology and, more significantly, in the Euclidean topology. [2] (See also Exercise 2.3.4.)

Exercise 3.2.2. Find a bijection between the set of all homogeneous polynomials in three variables of degree d and the set of all polynomials of degree at most d in two variables. (Hint: Set one of the variables to the constant 1.) Use this to show that the subspace topology induced on the affine patches $V \cap \mathbb{A}^2$ from the Zariski topology on a variety $V \subseteq \mathbb{P}^2$ is the same as the Zariski topology on the affine variety $V \cap \mathbb{A}^2$. Generalize to arbitrary dimension.

3.3 The Projective Closure of an Affine Variety

Let V be an affine algebraic variety in $\mathbb{A}^n$. We can visualize $\mathbb{A}^n$ as one of the coordinate leaves of $\mathbb{P}^n$, and hence, in a natural way as a dense open subset of $\mathbb{P}^n$. In this case, $\mathbb{P}^n$ is a natural "compactification" or "completion" of $\mathbb{A}^n$. Thus, we can also visualize V as a subset of $\mathbb{P}^n$.

Definition: Let V be an affine algebraic variety, considered with fixed embeddings $V \subseteq \mathbb{A}^n \subseteq \mathbb{P}^n$. The *projective closure* of V, denoted by $\overline{V}$, is the closure of V in the projective space $\mathbb{P}^n$. The closure may be computed either in the Zariski topology on $\mathbb{P}^n$, or in the Euclidean topology on $\mathbb{P}^n$; the result is the same, and both correspond to our intuitive idea of a closure.

Forming the projective closure of an affine algebraic variety gives a natural way to compactify any affine algebraic variety in the Euclidean topology. On the other hand, it is important to realize that the projective closure depends on the embedding of V in $\mathbb{P}^n$. In particular, isomorphic varieties can have non-isomorphic projective closures. See Exercise 3.4.4.

Example: Consider the parabola $V = \mathbb{V}(y - x^2) \subset \mathbb{A}^2 \subset \mathbb{P}^2$, illustrated in Figure 3.5. The variables x and y are the affine coordinates for V, that is, the coordinates of $\mathbb{A}^2$. In the projective plane we use homogeneous coordinates, x, y, and z, thinking of $\mathbb{A}^2$ as the open set U_z where z is nonzero (in the figure, $\mathbb{A}^2$ is identified with the plane $z = 1$ in $\mathbb{C}^3$).

Imagine the parabola in the projective space $\mathbb{P}^2$: Its points are the *lines* through the origin in $\mathbb{C}^3$ connecting to the points on the parabola in the plane $z = 1$. There is obviously one line "missing" from the cone over the parabola, namely, the line $x = z = 0$, the y-axis, where the two branches of the parabola come together. As a projective variety, our parabola ought

---

[2] A compactification of a topological space is a compact extension of the original space such that the original space is dense in the extension.

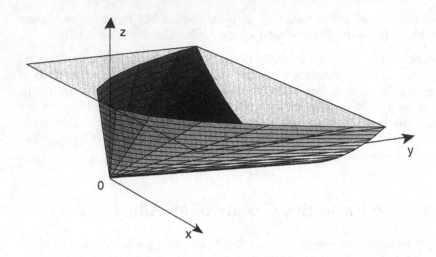

Figure 3.5. The projective closure of the parabola in $\mathbb{P}^2$

to be defined by the vanishing of $zy - x^2$ in $\mathbb{P}^2$. Note that it consists of our original parabola $y = x^2$ in the open set U_z, plus one "infinitely distant point" $[0 : 1 : 0]$.

If we know the ideal of an affine variety in $\mathbb{A}^n$, it is not difficult to find the ideal for its projective closure in $\mathbb{P}^n$. We now explain this process, called *homogenization*.

The polynomial $zy - x^2$ is called the *homogenization* of $y - x^2$. In general, a degree-d polynomial F in n variables is *homogenized* to become a degree-d homogeneous polynomial $\tilde{F}$ in $n + 1$ variables as follows: Decompose F into the sum of its homogeneous components of various degrees, $F = G_0 + G_1 + \ldots + G_d$, where G_i has degree i and some G_i's may be zero (but $G_d \neq 0$). Now, G_d already is homogeneous of degree d. The term $G_{d-1} \in \mathbb{C}[x_1, \ldots, x_n]$ is homogeneous of degree $d - 1$. We repair it, multiplying by a new variable x_0, to obtain a degree-d homogeneous polynomial $x_0 G_{d-1} \in \mathbb{C}[x_0, x_1, \ldots, x_n]$. Every term G_i can be made homogeneous of degree d by multiplying by x_0^{d-i}. The sum of the modified terms is the *homogenization* of F, a degree-d polynomial

$$\tilde{F} = x_0^d G_0 + x_0^{d-1} G_1 + \ldots + G_d.$$

Obviously, the restriction of $\tilde{F}$ to the hyperplane $x_0 = 1$ recovers our original polynomial F.

It is natural to guess that if $V = \mathbb{V}(F_1, \ldots, F_r) \subset \mathbb{A}^n$ is an algebraic variety, then the projective closure $\overline{V}$ of V in $\mathbb{P}^n$ might be defined by the ideal obtained by replacing each of the polynomials F_i by its homogenization $\tilde{F}_i$. This is the case, for example, with the parabola defined by the vanishing of $x^2 - y$, whose projective closure is defined by the vanishing of $x^2 - yz$. Unfortunately the situation is not so simple in general.

For example, let V be the subvariety of $\mathbb{A}^3$ defined by the vanishing of the two polynomials $y - x^2$ and $z - xy$. It is easy to see that V consists of the triples $\{(\lambda, \lambda^2, \lambda^3)\}$, that is, V is the twisted cubic we met earlier (see Exercise 1.2.3). Furthermore, the ideal generated by $y - x^2$ and $z - xy$ is the full radical ideal of polynomials vanishing on V. The homogenizations of these polynomials are $wy - x^2$ and $wz - xy$. However, the reader can check that the subvariety of $\mathbb{P}^3$ defined by these two polynomials consists of two components: The closure $\overline{V}$ of the twisted cubic plus the line defined by the vanishing of w and x. This extra line lies "at infinity" with respect to the affine chart U_w, which is why we did not see it before. This example shows that the closure $\overline{V}$ of V is not simply the variety in $\mathbb{P}^3$ defined by the homogenizations of a set of generators for $\mathbb{I}(V)$. Nonetheless, we have the following theorem.

Theorem: Let $V \subseteq \mathbb{A}^n \subseteq \mathbb{P}^n$ be an affine algebraic variety, and let $I \subseteq \mathbb{C}[x_1, \ldots, x_n]$ be the radical ideal of all polynomials vanishing on V. Then the ideal $\tilde{I}$ of $\mathbb{C}[x_0, \ldots, x_n]$ generated by the homogenization of *all* the elements of I is the radical homogeneous ideal of polynomials vanishing on the projective closure $\overline{V}$ in $\mathbb{P}^n$. The ideal $\tilde{I}$ is called the *homogenization* of the ideal I.

Of course, because $\tilde{I}$ is finitely generated, we know that I has some set of generators whose homogenizations produce $\tilde{I}$, but the preceding example shows that we must be careful about the choice of generators.

Proof of theorem: Let I be the radical ideal defining V in $\mathbb{A}^n$, and let $\tilde{I}$ be its homogenization. We need to show that $\overline{V} = \mathbb{V}(\tilde{I}) \subset \mathbb{P}^n$.

To show that $\overline{V} \subset \mathbb{V}(\tilde{I})$, it is enough to show that each G in $\tilde{I}$ vanishes on $\overline{V}$. For this, note that setting $x_0 = 1$, G is sent to a polynomial g in the ideal I. Thus G restricts to g on the open set U_0 where the homogeneous coordinate x_0 is nonzero, which is the affine patch of $\mathbb{P}^n$ containing V. This means that G vanishes on $V = \overline{V} \cap U_0$. So V, and hence its closure $\overline{V}$, is contained in the closed set $\mathbb{V}(\tilde{I})$.

For the converse we need to show that any polynomial vanishing on $\overline{V}$ is contained in $\tilde{I}$. If G is a homogeneous polynomial vanishing on $\overline{V}$, then G vanishes on $\overline{V} \cap U_0$. Thus the polynomial $G(1, x_1, \ldots, x_n) = g(x_1, \ldots, x_n)$ vanishes on V, and g is in I. By definition, the homogenization $\tilde{g}$ of g is in $\tilde{I}$. It is easy to check that $\tilde{g}x_0^t = G$ for some t, so G is in $\tilde{I}$ as well. We leave it to the reader to check that $\tilde{I}$ is radical in Exercise 3.3.3 below. $\square$

Exercise 3.3.1. Show that if V is an irreducible affine variety, then its projective closure $\overline{V}$ is also irreducible.

Exercise 3.3.2. Show that the twisted cubic curve V in $\mathbb{A}^3$ can also be defined by the polynomials $y - x^2$ and $z^2 - 2xyz + y^3$. Show that the two homogeneous polynomials obtained by homogenizing these polynomials define the projective closure $\overline{V}$ in $\mathbb{P}^3$, but that the ideal they generate is not radical.

Exercise 3.3.3. Show that the homogenization of a radical ideal is radical. (Hint: It suffices to show that if a power of a homogeneous polynomial is in the homogenization, then so is the polynomial.)

3.4 Morphisms of Projective Varieties

We turn our attention to morphisms between projective varieties, beginning with an example.

Consider the map

$$\mathbb{P}^1 \longrightarrow \mathbb{P}^2,$$
$$[s : t] \longmapsto [s^2 : st : t^2].$$

This map is well-defined. Indeed, if $[s : t] \in \mathbb{P}^1$, then

$$[s : t] = [\lambda s : \lambda t] \longmapsto [\lambda^2 s^2 : \lambda^2 st : \lambda^2 t^2] = [s^2 : st : t^2]$$

for any nonzero constant λ, so that the map does not depend on our choice of a representative for a point in $\mathbb{P}^1$. Also, at least one of the coordinates s^2 or t^2 of the image is nonzero.

Since $s^2 t^2 = (st)^2$, the image of this map lies on the curve $C = \mathbb{V}(xz - y^2)$ in $\mathbb{P}^2$. So the map defines a map from $\mathbb{P}^1$ to the curve C.

In the affine patch $\mathbb{A}^1 \cong U_t = \{[s : t] \mid t \neq 0\}$, denoting the coordinate $\frac{s}{t}$ by the letter u, this map can be written

$$\mathbb{A}^1 \longrightarrow \quad \mathbb{A}^2 \cong U_z = \{[x : y : z] \mid z \neq 0\}.$$
$$u \longmapsto \quad (u^2, u) \longmapsto [u^2 : u : 1].$$

The image of this map is the parabola $\mathbb{V}(x - y^2) \subseteq \mathbb{A}^2$. Similarly, in the other affine patch, $\mathbb{A}^1 \cong U_s = \{[s : t] \mid s \neq 0\}$, the map is described by

$$\mathbb{A}^1 \longrightarrow \quad \mathbb{A}^2 \cong U_x = \{[x : y : z] \mid x \neq 0\},$$
$$v \longmapsto \quad (v, v^2) \longmapsto [1 : v : v^2],$$

where $v = \frac{t}{s}$. Again, the image is a parabola in the plane.

The map $\mathbb{P}^1 \to C$ thus restricts, locally on the coordinate charts covering $\mathbb{P}^1$, to a morphism of affine algebraic varieties as defined in Chapter 1. This motivates the following definition.

Definition: Let $V \subseteq \mathbb{P}^n$ and $W \subseteq \mathbb{P}^m$ be projective algebraic varieties, and suppose that

$$V \xrightarrow{F} W$$

is a map from the set V to the set W. We say that F is a *morphism of projective varieties* if the following holds: For each $p \in V$, there exist homogeneous polynomials $F_0, \ldots, F_m \in \mathbb{C}[x_0, \ldots, x_n]$ such that for some nonempty open neighborhood $U \subseteq V$ of p, the map $U \xrightarrow{F|_U} W$ agrees with a polynomial map

$$
\begin{aligned}
U &\longrightarrow \mathbb{P}^m \\
q &\longmapsto [F_0(q) : F_1(q) : \cdots : F_m(q)].
\end{aligned}
$$

Of course, implicit in the definition is the fact that the homogeneous polynomials F_i all have the same degree. Otherwise, they will not describe a well-defined map to $\mathbb{P}^m$. Furthermore, the F_i must not vanish simultaneously on the open neighborhood U of p. It is important to realize that for different points p, it may be necessary to make a different choice of polynomials (and neighborhoods) to see that F is locally a polynomial map (we give an example below). The definition ensures that when a morphism of projective varieties is restricted to the coordinate charts, it defines a morphism of affine algebraic varieties.

By convention, when we say "open neighborhood" in the above definition, we mean a Zariski–open neighborhood. But those who have studied some complex analysis will quickly realize that it doesn't matter at all whether we use Zariski–open neighborhoods or open neighborhoods in the Euclidean topology. The main advantage of the Zariski topology is that it is available to us also in situations where the Euclidean topology is not, for instance if we work with algebraic varieties in projective space over some field other than $\mathbb{C}$ or $\mathbb{R}$.

Example: Let $C = \mathbb{V}(zx - y^2) \subseteq \mathbb{P}^2$ be a plane conic. Consider the map

$$
\begin{aligned}
C &\longrightarrow \mathbb{P}^1, \\
[x : y : z] &\longmapsto \begin{cases} [x : y] \text{ if } x \neq 0, \\ [y : z] \text{ if } z \neq 0. \end{cases}
\end{aligned}
$$

This map is defined at all points of C: Given a point of C, either its x or its z coordinate is nonzero, for if both were zero, then the relation $y^2 = xz$ would force y to be zero as well. To see that the map is well defined, note that if $[x : y : z]$ is a point of C with both its x and z coordinates (and hence its y coordinate) nonzero, then because the points of C satisfy $xz = y^2$, we see that

$$
[x : y] = [yx : y^2] = [xy : xz] = [y : z].
$$

This illustrates the important local nature of morphisms of projective varieties: The map $C \to \mathbb{P}^1$ is locally polynomial, but no single choice of polynomials will work for all points of C.

In the chart $U_x \cap C$, the morphism $C \to \mathbb{P}^1$ restricts to the projection of the parabola

$$(u, u^2) \longmapsto u$$

to the affine line. This map is the familiar *stereographic projection* of the conic curve C to the line $\mathbb{P}^1$. Our choice of affine coordinates so as to make C look like a parabola places the point from which we are projecting at infinity. For example, the point at infinity is $[0 : 0 : 1]$ when we consider the chart U_x.

Of course, whenever we have morphisms, we can also define isomorphisms. An *isomorphism* between projective varieties V and W is a morphism $V \xrightarrow{F} W$ such that there exists a morphism $W \xrightarrow{G} V$ that is inverse to F.

Example: The simplest example of an isomorphism is given by change of coordinates in $\mathbb{P}^n$. Explicitly, let $F_0, F_1, \ldots, F_n$ be linearly independent linear forms in $n+1$ variables. Then we have a morphism

$$\mathbb{P}^n \longrightarrow \mathbb{P}^n,$$
$$x \longmapsto [F_0(x) : F_1(x) : \ldots : F_n(x)].$$

The map is induced by the corresponding linear map of the vector space $\mathbb{C}^{n+1}$, and it is sometimes convenient to represent it by an $(n+1) \times (n+1)$ invertible matrix. The inverse morphism $\mathbb{P}^n \to \mathbb{P}^n$ is given by the inverse matrix.

Example: For a less trivial example, note that the conic curve $C \subseteq \mathbb{P}^2$ is isomorphic to $\mathbb{P}^1$. Indeed, the morphisms

$$\mathbb{P}^1 \longrightarrow C,$$
$$[s : t] \longmapsto [s^2 : st : t^2],$$

and

$$C \longrightarrow \mathbb{P}^1$$
$$[x : y : z] \longmapsto \begin{cases} [x : y] \text{ if } x \neq 0, \\ [y : z] \text{ if } z \neq 0, \end{cases}$$

are easily checked to be inverse to each other.

Here we see a big difference between the theory of affine and projective varieties. Two affine algebraic varieties are isomorphic if and only if their coordinate rings are isomorphic as $\mathbb{C}$-algebras. However, the corresponding statement is quite false for projective varieties. In the example above, C and $\mathbb{P}^1$ are isomorphic, but their homogeneous coordinate rings,

$$\frac{\mathbb{C}[x, y, z]}{(xz - y^2)} \quad \text{and} \quad \mathbb{C}[s, t],$$

are not isomorphic as $\mathbb{C}$-algebras. Indeed, an isomorphism of the $\mathbb{C}$-algebras would correspond to an isomorphism between the *affine cones* over C and $\mathbb{P}^1$. These are the varieties

$$\mathbb{V}(xz - y^2) \subseteq \mathbb{A}^3 \quad \text{and} \quad \mathbb{A}^2.$$

It is intuitively clear that these varieties ought not be isomorphic, because the former looks like a cone, with a singular point at the origin, while the affine plane has no singularities.

A projective variety is determined, up to isomorphism, by its homogeneous coordinate ring, but not conversely. However, there is a stronger type of isomorphism between projective varieties that does guarantee an isomorphism between the corresponding homogeneous coordinate rings.

Definition: Two subvarieties of $\mathbb{P}^n$ are said to be *projectively equivalent* if there exists a (linear) change of coordinates on $\mathbb{P}^n$ that defines an isomorphism between them.

For example, the lines $\mathbb{V}(x)$ and $\mathbb{V}(y)$ are projectively equivalent subvarieties of $\mathbb{P}^2$. The necessary change of coordinates is

$$\mathbb{P}^2 \longrightarrow \mathbb{P}^2,$$
$$[x : y : z] \longmapsto [y : x : z].$$

Exercise 3.4.1. Find the matrix that describes the projective equivalence above.

Exercise 3.4.2. Show that the homogeneous coordinate rings of projectively equivalent varieties are isomorphic.

Exercise 3.4.3. Find an example of two plane projective curves that are isomorphic but not projectively equivalent.

Exercise 3.4.4. Show that the affine varieties $\mathbb{V}(x)$ and $\mathbb{V}(x - y^4 - z^4)$ in $\mathbb{A}^3$ are isomorphic but that their projective closures in $\mathbb{P}^3$ are not. (Hint: If you find it difficult to show rigorously that the projective closures are not isomorphic, try again after reading about smoothness in section 6.2.)

3.5 Automorphisms of Projective Space

We now determine all isomorphisms from projective space to itself, the *automorphisms* of $\mathbb{P}^n$.

If you have seen some complex analysis, you know that the conformal mappings from the Riemann sphere to itself are exactly the Möbius transformations, also known as the fractional linear transformations. Because every automorphism of $\mathbb{P}^1$ is locally polynomial, and hence holomorphic, it follows that every automorphism of the projective variety $\mathbb{P}^1$ is a conformal map of the Riemann sphere to itself. On the other hand, it is easy to see that a Möbius transformation is a morphism of the projective variety $\mathbb{P}^1$, as we will soon explain below. In other words, the automorphisms of the projective line are exactly the Möbius transformations. This idea generalizes to projective spaces of arbitrary dimension.

Recall that an invertible matrix $g \in \mathbf{GL}(n+1, \mathbb{C})$ defines a vector space map

$$\mathbb{C}^{n+1} \xrightarrow{g} \mathbb{C}^{n+1},$$
$$(x_0, \ldots, x_n) \longmapsto g(x_0, \ldots, x_n),$$

which determines an automorphism of projective space

$$\mathbb{P}^n \xrightarrow{g} \mathbb{P}^n,$$
$$[x_0 : \cdots : x_n] \longmapsto g([x_0 : \cdots : x_n]).$$

This automorphism is simply a linear coordinate change in $\mathbb{P}^n$, and its inverse automorphism is given by the inverse matrix g^{-1}. Because the coordinates of $\mathbb{P}^n$ are defined only up to scalar multiple, the matrices g and λg define the same isomorphism, for any nonzero scalar λ in $\mathbb{C}$. On the other hand, matrices differing in some other way produce different isomorphisms.

It turns out that every automorphism of projective space has this form, that is, the only automorphisms of $\mathbb{P}^n$ are linear changes of coordinates. It is a tedious job to prove this by elementary calculations, but those familiar with complex analysis can easily show that every biholomorphic automorphism of $\mathbb{P}^n$ is a linear change of coordinates. There is an elegant algebraic proof, but it requires tools not presented here (see [20, page 151]).

A fancy way to say that every automorphism of $\mathbb{P}^n$ is a linear coordinate change is to say that the automorphism group of $\mathbb{P}^n$ is

$$\mathbf{PGL}(n+1, \mathbb{C}) = \mathbf{GL}(n+1, \mathbb{C}) / \mathbb{C}^*,$$

the group of invertible $(n+1) \times (n+1)$ matrices modulo the subgroup of nonzero scalar matrices.

We easily check that this result generalizes the familiar one dimensional case, that is, that $\mathbf{PGL}(2, \mathbb{C})$ is the group of Möbius transformations. Indeed, any element of $\mathbf{PGL}(2, \mathbb{C})$ determines a map

$$\mathbb{P}^1 \to \mathbb{P}^1,$$
$$[z : w] \longmapsto [az + bw : cz + dw],$$

in other words, a Möbius map,

$$[z : 1] \longmapsto \left[\frac{az + b}{cz + d} : 1 \right].$$

Conversely, a Möbius transformation of the form $z \mapsto (az + b)/(cz + d)$ gives rise to the automorphism of $\mathbb{P}^1$ given by $[z : w] \mapsto [az + bw : cz + dw]$.

We can now state the definition of projective equivalence given in Section 3.4 in a slightly different form. Two subvarieties of projective space are projectively equivalent if and only if they differ by an automorphism of the ambient projective space. That is, the equivalence classes of projectively equivalent subvarieties of $\mathbb{P}^n$ are precisely the orbits of the natural action of $\mathbf{PGL}(n+1, \mathbb{C})$ on the set of all projective subvarieties of $\mathbb{P}^n$.

Our discussion indicates that there is no difference between the algebraic automorphisms and the biholomorphic automorphisms of $\mathbb{P}^n$. In fact, there is no difference between the algebraic and the complex analytic categories in much more general settings:

Chow's Theorem: Every compact complex manifold embedded in $\mathbb{P}^n$ is the common zero set of some homogeneous polynomials $F_1, \ldots, F_r$. So every compact complex submanifold of projective space is a projective variety. Furthermore, every holomorphic mapping between such manifolds is a morphism of varieties.

We do not prove this here; instead, we refer the reader to [37, part III]. Chow's theorem fails without the compactness hypotheses. Serre vastly generalized Chow's theorem in his famous "GAGA" paper [36].

There is a partial converse to Chow's theorem in the one-dimensional case. Every complex one-dimensional compact manifold (that is, every Riemann surface) can be embedded as a complex manifold in projective space, and hence is defined by polynomial equations. Thus, a smooth[3] projective one-dimensional algebraic variety is "the same as" a compact Riemann surface: There is a unique way to define a complex structure on a smooth projective curve, and a unique way to define the structure of a projective variety on each compact Riemann surface.

Exercise 3.5.1. Let F and $G \in \mathbb{C}[x, y, z]$ be two irreducible homogeneous quadratic polynomials. Show that there exists an automorphism of $\mathbb{P}^2$ mapping $\mathbb{V}(F)$ isomorphically onto $\mathbb{V}(G)$. This shows that there exists only one (nondegenerate) projective conic up to a linear change of coordinates (Hint: An irreducible homogeneous quadratic polynomial defines a nondegenerate quadratic form on $\mathbb{C}^3$).

Exercise 3.5.2. Show that up to affine change of coordinates in the affine plane, there exist exactly two nonisomorphic affine (nondegenerate) conic plane curves. That is, the zero set of any irreducible quadratic polynomial $F \in \mathbb{C}[x, y]$ is—up to a linear change of coordinates—either a parabola $\mathbb{V}(y - x^2)$ or a hyperbola $\mathbb{V}(xy - 1)$, but the parabola and hyperbola are nonisomorphic.

Remark: After completing the first Exercise above, you might hope that

---

[3]Smoothness of an algebraic variety will be defined precisely in Section 6.2, here we rely on the reader's intuitive idea.

there is only one plane cubic up to isomorphism. However, there is a continuum of nonisomorphic plane cubics, the elliptic curves, parametrized by $\mathbb{A}^1$ using the so-called j-invariant (see [20, Chapter IV, Section 4]).

4
Quasi-Projective Varieties

4.1 Quasi-Projective Varieties

We have developed the theory of affine and projective varieties separately. We now introduce the concept of a quasi-projective variety, a term that encompasses both cases. More than just a convenience, the notion of a quasi-projective variety will eventually allow us to think of an algebraic variety as an intrinsically defined geometric object, free from any particular embedding in affine or projective space.

Definition: A *quasi-projective variety* is a locally closed subset of $\mathbb{P}^n$, considered with the Zariski topology induced from $\mathbb{P}^n$. Recall that a *locally closed set* of any topological space is a closed subset of an open subspace, in other words, an intersection of an open set and a closed set.

The class of quasi-projective varieties includes all projective varieties, all affine varieties, and all Zariski open subsets of these. The set of quasi-projective varieties is closed under taking open or closed subsets. For brevity, we often say "variety," instead of quasi-projective variety.

We can define a morphism between quasi-projective varieties in exactly the same way we defined a morphism between projective varieties. To be precise, if $V \subseteq \mathbb{P}^n$ and $W \subseteq \mathbb{P}^m$ are quasi-projective varieties, then a *morphism* $V \xrightarrow{F} W$ is a map such that for each $p \in V$, there exist homogeneous polynomials $F_0, \ldots, F_m$ in $n+1$ variables such that the "map"

$$V \to \mathbb{P}^m,$$
$$q \mapsto [F_0(q) : \cdots : F_m(q)],$$

is well-defined at p and agrees with F on some noncmpty open set containing p.

One disadvantage of this definition is that it refers to a specific embedding of the quasi-projective varieties in projective space. We will give an equivalent definition in Section 4.3 that avoids this inconvenience.

Example: Let $U = \mathbb{A}^1 \smallsetminus \{0\}$ and let $V = \mathbb{V}(xy - 1) \subseteq \mathbb{A}^2$. Both U and V are quasi-projective varieties, and we have a well-defined map

$$U \xrightarrow{F} V,$$
$$t \longmapsto \left(t, \tfrac{1}{t}\right).$$

We claim that this is a morphism of quasi-projective varieties. To see this, we identify U with a locally closed (in fact, open) subset of $\mathbb{P}^1$ by the map $t \mapsto [t : 1]$. Likewise, we identify V with a locally closed subset of $\mathbb{P}^2$ defined by $xy - z^2 = 0$ and $z \neq 0$ via the map $(x, y) \mapsto [x : y : 1]$. It is easy to check that the map F agrees everywhere on U with the morphism

$$\mathbb{P}^1 \xrightarrow{\tilde{F}} \mathbb{P}^2,$$
$$[a : b] \longmapsto [a^2 : b^2 : ab].$$

Indeed, on U neither a nor b is zero, so setting $t = \frac{a}{b}$, we see that $\tilde{F}$ sends $[t : 1]$ to

$$[t^2 : 1 : t] = \left[t : \frac{1}{t} : 1\right].$$

The image of this map is precisely $V \subseteq \mathbb{P}^2$, and the map obviously agrees with the original map F.

By enlarging our world to include all quasi-projective varieties, we gain flexibility. However, we must redefine the concept of an affine variety.

Definition: A quasi-projective variety is said to be *affine* if it is isomorphic as a quasi-projective variety to some affine algebraic variety. From now on, when we want to refer to the original definition of an "affine variety in $\mathbb{A}^n$," we will call it a Zariski-closed subset of $\mathbb{A}^n$.

Thus an affine algebraic variety is a variety that *can be* embedded in affine space as a Zariski-closed subset. With this refined definition of an affine variety, we emphasize the intrinsic nature of the variety, rather than the extrinsic feature of a particular embedding in affine space as in Section 1.1. This change of definition actually enlarges the class of affine varieties, as the next example shows.

Example: The open set $U = \mathbb{A}^1 \smallsetminus \{0\}$ in $\mathbb{A}^1$ is an affine variety. The projection map $V = \mathbb{V}(xy - 1) \xrightarrow{G} U$ given by $G(x, y) = x$ is a morphism of varieties (see Figure 4.1). Letting F denote the morphism $U \to V$ defined above, we see that $F \circ G$ is the identity map on V and $G \circ F$ is the identity map on U. So $\mathbb{A}^1 \smallsetminus \{0\}$ is a quasi-projective variety that is isomorphic to the Zariski-closed set $\mathbb{V}(xy - 1)$ in $\mathbb{A}^2$. Accordingly, we now call $\mathbb{A}^1 \smallsetminus \{0\}$ an affine variety.

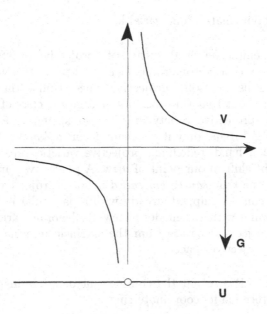

Figure 4.1. An isomorphism of quasi-projective varieties

The *coordinate ring of an affine variety* W is defined to be the coordinate ring of any closed subvariety of affine space isomorphic to W. To be precise, fix an isomorphism of quasi-projective varieties

$$W \xrightarrow{F} V,$$

where V is a Zariski-closed set of some affine space. The coordinate ring $\mathbb{C}[W]$ on W is defined to be the ring of all functions $W \to \mathbb{C}$ that are pullbacks of functions f in $\mathbb{C}[V]$. It is not difficult to check that this is well-defined, that is, it does not depend on the choice of V or on the choice of the isomorphism F between W and V. In checking this fact, the reader will need to verify that if two Zariski closed subvarieties of affine spaces are isomorphic as quasi-projective varieties, then they are also isomorphic as affine algebraic varieties as defined in Section 1.3. That is, any isomorphism of quasi-projective varieties that are Zariski-closed subsets of some ambient affine spaces is in fact the restriction of a polynomial map on the ambient spaces. This fact is not completely trivial, but will be easier after reading Section 3 on regular functions, where related facts are proved.

In the example above, the coordinate ring of $\mathbb{A}^1 \smallsetminus \{0\}$ is isomorphic to

$$\mathbb{C}[V] = \frac{\mathbb{C}[x,y]}{(xy-1)} \cong \mathbb{C}\left[x, \frac{1}{x}\right],$$

the Laurent polynomials in one variable.

We can also reinterpret the definition of a projective variety as any quasi-projective variety that is isomorphic, as a quasi-projective variety, to some Zariski-closed subset of some projective space. Unlike the case of affine varieties, it turns out that this does not enlarge the class of projective varieties: A quasi-projective variety in $\mathbb{P}^n$ is isomorphic to a Zariski-closed subset of some $\mathbb{P}^m$ if and only if it already forms a Zariski-closed subset of $\mathbb{P}^n$. On the other hand, redefining projective varieties this way does lead to an important shift in our point of view. A projective variety should be interpreted as one that *can be embedded* in some projective space, rather than one that comes equipped already with a particular embedding. This is more in keeping with the modern view of algebraic varieties as intrinsic geometric objects, separate from the extrinsic information of a fixed embedding in projective space.

Exercise 4.1.1. Prove that the complement of a line in $\mathbb{A}^2$ is an affine variety and determine its coordinate ring.

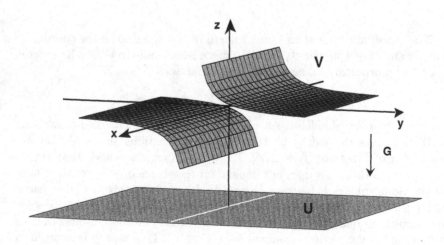

Figure 4.2. The complement of a line in $\mathbb{A}^2$ is affine

4.2   A Basis for the Zariski Topology

The Zariski topology for any quasi-projective variety has a basis of open affine sets. This important fact allows us to think of every quasi-projective variety as "locally affine" in the same way that a manifold is "locally Euclidean": Each point p of the variety has an open neighborhood that is an affine algebraic variety.

To see this, first we observe that the complement of any hypersurface in an affine variety is again an affine variety. More precisely, if V is a Zariski-closed subset of affine space $\mathbb{A}^n$, and f is any function in the coordinate ring $\mathbb{C}[V]$ of V, then the open set

$$U = V \smallsetminus \mathbb{V}(f)$$

is again an *affine* algebraic variety (though not usually a closed subvariety of V, and hence not a Zariski-closed set of the ambient $\mathbb{A}^n$).

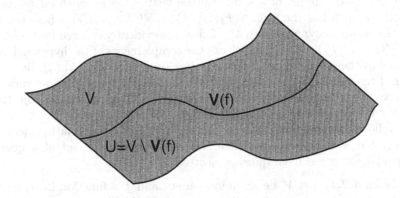

Figure 4.3. Complement of a Hypersurface

Indeed, think of U as a subset of $\mathbb{A}^n$ and consider the map

$$U \xrightarrow{F} \mathbb{A}^{n+1}$$
$$(x_1, \dots, x_n) \longmapsto \left(x_1, \dots, x_n, \tfrac{1}{f(x_1, \dots, x_n)} \right).$$

Because f does not vanish on U, this map is well-defined. Furthermore, if $x_1, \dots, x_n, z$ denote the coordinates for $\mathbb{A}^{n+1}$, then the original defining polynomials $F_1(x_1, \dots, x_n), \dots, F_r(x_1, \dots, x_n)$ for V in $\mathbb{A}^n$ all vanish at the image points of F, as does the polynomial $zf(x_1, \dots, x_n) - 1$. In other words, the image of F lies in the Zariski-closed subset of $\mathbb{A}^{n+1}$ defined by $W = \mathbb{V}(F_1, \dots, F_r, zf - 1)$.

The reader can easily check that this map of sets

$$U \to \mathbb{V}(F_1, \dots, F_r, zf - 1) \subset \mathbb{A}^{n+1}$$

is an *isomorphism* of quasi-projective varieties, following exactly the same argument as in the case of the complement of a point in $\mathbb{A}^1$. This shows that the open subset U of the affine variety V defined as the complement of the vanishing set of a single polynomial function f on V is again an *affine* variety. By definition, the coordinate ring of U is isomorphic to $\mathbb{C}[W] = \frac{\mathbb{C}[x_1,\ldots,x_n,z]}{(F_1,\ldots,F_r,zf-1)}$. The reader will easily verify that $\mathbb{C}[W] \cong \mathbb{C}[V][\frac{1}{f}]$.

Caution: Not all open sets of affine or projective space are affine. Indeed, the punctured plane, $\mathbb{A}^2 \smallsetminus \{0\}$, is not affine. We do not yet have the tools to give a concise proof that this is not affine, but we will return to this example in the next section.

We can now see why every quasi-projective variety V has a basis of open affine sets. As we have seen, thinking of V as a subset of projective space $\mathbb{P}^n$, V is the union of its intersection with each of the coordinate charts U_i in $\mathbb{P}^n$. Now, each $V \cap U_i$ is a quasi-projective variety in $U_i \cong \mathbb{A}^n$, and because a locally closed subset of a space can be written as an open subset of a closed set, each has the form $\mathbb{V}(F_1,\ldots,F_s) \smallsetminus \mathbb{V}(G_1,\ldots,G_t)$, where the F_j and the G_j are polynomials on $\mathbb{A}^n_i$. This set is evidently covered by the open sets $\mathbb{V}(F_1,\ldots,F_s) \smallsetminus \mathbb{V}(G_j)$, which are the complements of the hypersurfaces defined by the restrictions of the G_j to the closed set $\mathbb{V}(F_1,\ldots,F_s)$. Because each of these sets is the complement of a hypersurface in an affine variety, each is affine; furthermore, each is open in V. Thus the quasi-projective variety V has a cover by open affine sets.

It follows immediately that the Zariski topology of any quasi-projective variety has a basis of open affine sets, since each open set of a quasi-projective variety is itself quasi-projective.

Exercise 4.2.1. Let V be an affine variety and f a function in its coordinate ring. Show that if f vanishes nowhere on V, then f is invertible in $\mathbb{C}[V]$.

Exercise 4.2.2. Find an open affine cover of the punctured plane $\mathbb{A}^2 \smallsetminus \{(0,0)\}$.

Exercise 4.2.3. Show that the set $\mathbf{GL(n,\mathbb{C})}$ of invertible $n \times n$ matrices has the structure of an affine algebraic variety (see Section 1.1).

4.3 Regular Functions

Regular functions on a quasi-projective variety are a natural generalization of polynomial functions on an affine variety.

Behind the definition of a regular function is the idea that quasi-projective varieties are like manifolds in many respects. Whereas manifolds look locally like the Euclidean space $\mathbb{R}^n$, varieties look locally like affine varieties. The existence of a basis of open affine sets means that we can think of a variety as a union of affine varieties, and so we define a regular function locally—its restriction to each affine patch should be a polynomial function.

First consider a Zariski-closed subset of $\mathbb{A}^n$, say V. Given any two functions f and g in the coordinate ring $\mathbb{C}[V]$, the rational expression $\frac{f}{g}$ can be thought of as a locally defined function: $\frac{f}{g}$ is well-defined on the open set $V \smallsetminus \mathbb{V}(g)$. As we have seen, this open set is isomorphic to an affine variety with coordinate ring $\frac{\mathbb{C}[V][z]}{(zg-1)}$, a Zariski-closed set in $\mathbb{A}^{n+1}$; the corresponding isomorphism is analogous to a chart map for a manifold that sends an open set in the manifold to an open set in Euclidean space. On our chart $V \setminus \mathbb{V}(g)$, the function $\frac{1}{g}$ is identified with the polynomial function z on $\mathbb{A}^{n+1}$, and the function $\frac{f}{g}$ is identified with the polynomial function zf on $\mathbb{A}^{n+1}$. In this sense each rational function on V is a polynomial function on some open subset of V.

Now we define a regular function on an affine variety that is not necessarily a closed subset of $\mathbb{A}^n$.

Definition: Let U be any open set of an affine variety V. A complex-valued function $U \xrightarrow{f} \mathbb{C}$ is *regular at a point* $p \in U$ if there exist functions g and h in $\mathbb{C}[V]$ such that $h(p) \neq 0$ and f agrees with the function $\frac{g}{h}$ in some neighborhood of p. The function f is *regular on U* if it is regular at every point of U. The set of all regular functions on U is denoted by $\mathcal{O}_V(U)$.

If V is a Zariski-closed subset of $\mathbb{A}^n$, then obviously each element $g \in \mathbb{C}[V]$ in its coordinate ring defines a regular function $V \to \mathbb{C}$. In other words, there is a natural inclusion $\mathbb{C}[V] \subset \mathcal{O}_V(V)$. Actually, we will soon see that a regular function $f \in \mathcal{O}_V(V)$ must be the restriction of some polynomial function on $\mathbb{A}^n$ to V, so that $\mathbb{C}[V] = \mathcal{O}_V(V)$. Before giving the proof, we discuss some examples.

Examples: (1) The slope function

$$U = \mathbb{A}^2 \smallsetminus \mathbb{V}(x) \xrightarrow{f} \mathbb{C},$$
$$(x, y) \longmapsto \tfrac{y}{x},$$

is regular on the set U.

(2) Projection from a point in the plane defines a regular function as follows. Take any $p \in \mathbb{A}^2$. Choose a line $\ell \subset \mathbb{A}^2$, not passing through p, and identify ℓ with the complex line $\mathbb{A}^1$. Let ℓ' be a line, parallel to ℓ and passing through p.
Define a function

$$\mathbb{A}^2 \smallsetminus \ell' \xrightarrow{\varphi} \ell = \mathbb{C},$$
$$q \longmapsto \text{the only element in } \overline{qp} \cap \ell.$$

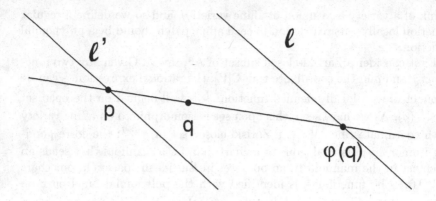

Figure 4.4. Projection from a point

The reader will easily verify the regularity of ϕ on the open set $\mathbb{A}^2 \setminus \ell'$ by expressing it as a rational function of the coordinates of $\mathbb{A}^2$; see Exercise 4.3.2.

Theorem: Let V be an irreducible Zariski closed subset of $\mathbb{A}^n$. Then every regular function $V \xrightarrow{g} \mathbb{C}$ is the restriction of some polynomial $\mathbb{A}^n \xrightarrow{f} \mathbb{C}$ to V. In other words, $\mathcal{O}_V(V) = \mathbb{C}[V]$.

Proof: We have already seen that $\mathbb{C}[V] \subseteq \mathcal{O}_V(V)$.

Now let $g \in \mathcal{O}_V(V)$ be any regular function on V. By definition, for each $p \in V$, we can find an open neighborhood U_p of p such that g agrees with some rational function $\frac{h_p}{k_p}$, where h_p and k_p are elements of $\mathbb{C}[V]$ with $k_p(p) \neq 0$.

Because the affine open sets of the type $U_F = V \setminus \mathbb{V}(F)$ form a basis for the Zariski topology on V, we can assume that each of the open sets U_p is of the form U_F for some F (depending on p). Furthermore, because the Zariski topology on V is compact, there is a finite subcover of the cover $\{U_p\}_{p \in V}$ of V consisting of sets of the form $U_{F_1}, \dots, U_{F_t}$ where g agrees with $\frac{h_i}{k_i}$ on $U_{F_i} = V \setminus \mathbb{V}(F_i)$. Thus

$$g|_{U_{F_i}} = \frac{h_i}{k_i} \qquad \text{for each } i = 1, \dots, t.$$

Because the open sets $\{U_{F_i}\}_{i=1}^t$ cover V, the polynomials k_i cannot all simultaneously vanish on V. Thus $\mathbb{V}(k_1, \dots, k_t) = \emptyset$, and Hilbert's Nullstellensatz implies that the ideal generated by the k_i's must be the unit ideal of $\mathbb{C}[V]$. This means that we can write

$$1 = \sum_{j=1}^t \ell_j k_j$$

for some polynomial functions $\ell_j \in \mathbb{C}[V]$. Then, in every U_{F_i} we have

$$g = 1 \cdot g = \sum_{j=1}^{t} \ell_j k_j \frac{h_i}{k_i} \in \mathbb{C}[V].$$

Set

$$f = \sum_{j=1}^{t} h_j \ell_j \in \mathbb{C}[V].$$

We claim that $f = g$ as functions on V, and this will complete the proof. To see this, first note that because g restricts to $\frac{h_i}{k_i}$ on U_{F_i} for all i, we must have that

$$\frac{h_i}{k_i} = \frac{h_j}{k_j}$$

on the dense open set $U_{F_i} \cap U_{F_j}$, whence the polynomials $h_i k_j$ and $h_j k_i$ agree as functions on V, for all pairs i, j. Thus on each U_{F_i},

$$g = 1 \cdot g = \left(\sum_{j=1}^{t} \ell_j k_j \right) \cdot \left(\frac{h_i}{k_i} \right) = \sum_{j=1}^{t} \ell_j (h_j k_i) \frac{1}{k_i} = \sum_{j=1}^{t} \ell_j h_j = f.$$

Because g agrees with f on each open set in the cover U_{F_i} of V, we conclude that $g = f \in \mathbb{C}[V]$. $\qquad\square$

The theorem is true even when V is not irreducible, but the proof in this more general case requires more technical algebra.[1] The problem is that while $\frac{h_i}{k_i} = \frac{h_j}{k_j}$ on $U_{F_i} \cap U_{F_j}$, the set $U_{F_i} \cap U_{F_j}$ may not be dense, so this does not imply that $h_i k_j = h_j k_i$ in the general case.

The preceding theorem is important because it ensures that our *local* definition of a regular function on an affine variety produces precisely the same functions we have already adopted as the natural ones to look at when studying affine varieties, namely, the restrictions of polynomial maps on the ambient affine spaces. We can now confidently state the definition of a regular function on an arbitrary quasi-projective variety.

Definition: Let U be an open subset of a quasi-projective variety V. A complex-valued function $U \xrightarrow{f} \mathbb{C}$ on U is *regular at* $p \in U$ if there exists some affine open set containing p on which f is regular at p. A function is regular on U if it is regular at every point of U. The set of all regular functions on U is denoted by $\mathcal{O}_V(U)$.

If the quasi-projective variety V happens to be affine, this definition agrees with the definition of a regular function on an affine variety.

---

[1]See [20, Proposition 2.2 on page 71].

Fix a quasi-projective variety V and consider the set $\mathcal{O}_V(U)$ for each open set U in V. The local nature of regular functions can be summarized in the following properties.

1. Every $\mathcal{O}_V(U)$ is a ring (in fact, a $\mathbb{C}$-algebra) with respect to pointwise addition and multiplication.

2. If a function is regular in U, it will also be regular in any open subset of U, and if $U_1 \subset U_2$ are open subsets of V, then restriction defines a natural ring homomorphism $\mathcal{O}_V(U_2) \to \mathcal{O}_V(U_1)$.

3. If two regular functions f_1 and f_2, defined on $U_1 \subset V$ and $U_2 \subset V$ respectively, agree on the intersection $U_1 \cap U_2$, they uniquely define a function f on the union $U_1 \cup U_2$. The function f is regular on the union, and f_1 and f_2 are the restrictions of f to the original sets. The same can be done for more than two functions, in fact for infinitely many, provided that the functions f_i agree on all pairwise intersections of the sets U_i.

4. Regularity of functions is preserved by pullbacks with respect to morphisms. More precisely, if $V \xrightarrow{F} W$ is a morphism of quasi-projective varieties and $U \subset W$ is an open set, then for any $f \in \mathcal{O}_W(U)$,

$$f \circ F \in \mathcal{O}_V(F^{-1}(U)).$$

A formal way to phrase the first three properties above is that every quasi-projective variety comes equipped with a natural *sheaf* of $\mathbb{C}$-algebras, denoted by $\mathcal{O}_V$ and called the *structure sheaf of V*. Property 4 says that every morphism of algebraic varieties induces a natural morphism of the corresponding sheaves of regular functions. The appendix contains more about sheaves and the use of sheaves to define *schemes* and *abstract algebraic varieties*.

The definition of a morphism of quasi-projective varieties can also be rephrased in a local way, using regular functions.

Definition: A map $V \xrightarrow{\phi} W$ of quasi-projective varieties is a *morphism* if for all $p \in V$, there exist open affine neighborhoods U of p and U' of $\phi(p)$ such that $\phi(U) \subseteq U'$ and $\phi|_U$ agrees with a map of affine varieties as defined in Section 1.3, that is, $\phi|_U$ should be given by a set of regular functions in the coordidnates of U.

The reader can check that this agrees with our previous definition of a morphism of quasi-projective varieties. The advantage of this new definition is that it does not require us first to embed the quasi-projective variety into projective space in order to describe the morphism.

Exercise 4.3.1. Show that the projection map described in the example is regular. (Hint: Choose (or change) coordinates for the plane so that the point p is the origin, ℓ' is the y-axis, and ℓ is the line $x = 1$.)

Exercise 4.3.2. Show that the ring $\mathcal{O}_V(U)$ of regular functions on the punctured plane $\mathbb{A}^2 \smallsetminus \{(0,0)\}$ is the polynomial ring $\mathbb{C}[x, y]$. Conclude that this quasi-projective variety is not affine.

5
Classical Constructions

5.1 Veronese Maps

Veronese maps provide an important example of morphisms of quasi-projective varieties. A Veronese map embeds a projective space $\mathbb{P}^n$ as a subvariety of some higher-dimensional projective space in a nontrivial way.

Consider the set of *all* homogeneous polynomials of fixed degree d in the polynomial ring $\mathbb{C}[x_0, \ldots, x_n]$. This is a finite-dimensional $\mathbb{C}$-vector space, and the $\binom{d+n}{d}$ monomials of the form $x_0^{d_0} \cdots x_n^{d_n}$ with $\sum d_i = d$ form a basis.

Definition: The dth *Veronese mapping* of $\mathbb{P}^n$ is the morphism

$$\mathbb{P}^n \xrightarrow{\nu_d} \mathbb{P}^m,$$
$$[x_0 : \cdots : x_n] \longmapsto \underbrace{[x_0^d : x_0^{d-1}x_1 : \cdots : x_n^d]}_{\text{all monomials of degree } d},$$

where $m = \binom{d+n}{d} - 1$.

This is a well-defined morphism of projective varieties, because the defining polynomials all have the same degree and do not all simultaneously vanish at any point of $\mathbb{P}^n$.

Proposition: The image of the Veronese mapping ν_d is a closed subvariety of $\mathbb{P}^m$, and ν_d is an isomorphism from $\mathbb{P}^n$ onto this subvariety. In other words, the Veronese mapping is an embedding of algebraic varieties.

Proof: We describe the inverse map. Let $W \subseteq \mathbb{P}^m$ be the image of ν_d. Note that the homogeneous coordinates of $\mathbb{P}^m$ are indexed by the degree-d monomials in $(n+1)$ variables; we can write them as z_I for $I = (i_0, \ldots, i_n) \in \mathbb{N}^{n+1}$ with $i_0 + \ldots + i_n = d$.

At each point of W, at least one of the coordinates indexed by the monomials $x_0^d, \ldots, x_n^d$ must be nonzero. Let $U_i \subset W$ be the subset of W where the coordinate indexed by x_i^d is nonzero. The sets $U_0, \ldots, U_n$ cover W and we can define a map

$$U_i \longrightarrow \mathbb{P}^n,$$
$$z \longmapsto [z_{(1,0,\ldots,d-1,\ldots,0)} : z_{(0,1,0,\ldots,d-1,0,\ldots,0)} : \cdots : z_{(0,\ldots,d-1,0,\ldots,1)}],$$

for $z \in U_i$. That is, we send z to the $(n+1)$-tuple of its coordinates indexed by $x_0 x_i^{d-1}, \ldots, x_n x_i^{d-1}$. These maps agree on the overlaps $U_i \cap U_j$, so these maps patch together to define a map $W \to \mathbb{P}^n$. The composition $\mathbb{P}^n \to W \to \mathbb{P}^n$ looks like $[x_0 : \cdots : x_n] \mapsto \nu_d(x) \mapsto [x_0 x_i^{d-1} : \cdots : x_n x_i^{d-1}] = [x_0 : \cdots : x_n]$, the identity map. Equally easily, one checks that $W \to \mathbb{P}^n \to W$ is the identity map on W. This shows that there are invertible polynomial maps defining a bijection between $\mathbb{P}^n$ and $\nu_d(\mathbb{P}^n)$, so the proof is essentially complete. We need only verify that the image set $\nu_d(\mathbb{P}^n)$ really is a subvariety of $\mathbb{P}^m$. This is easy to do after considering a few examples to get an idea what is happening. $\square$

Example: We begin with the case $n = 1$ and $d = 2$. Then the Veronese map is

$$\mathbb{P}^1 \xrightarrow{\nu_2} \mathbb{P}^2,$$
$$[s : t] \longmapsto [s^2 : st : t^2].$$

Its image is the conic curve $\mathbb{V}(xz - y^2)$ in $\mathbb{P}^2$. We have already seen in Section 3.4 that this map defines an isomorphism onto its image. So the Veronese map ν_2 induces an isomorphism between $\mathbb{P}^1$ and a conic in $\mathbb{P}^2$.

Example: The Veronese map

$$\mathbb{P}^1 \xrightarrow{\nu_3} \mathbb{P}^3,$$
$$[s : t] \longmapsto [\underbrace{s^3}_{x} : \underbrace{s^2 t}_{y} : \underbrace{st^2}_{z} : \underbrace{t^3}_{w}],$$

is also an isomorphism onto its image. Its image is called the *rational normal curve* of degree 3. This is the projective closure of the twisted cubic we met in Section 3.3. (Since the meaning is usually clear from the context, we will refer to both the affine and projective curves as the twisted cubic.) It can be checked that the image of ν_3 is the projective variety defined by the polynomials $xw - yz, y^2 - xz$, and $wy - z^2$, where x, y, z, w are the homogeneous coordinates on $\mathbb{P}^3$.

The other Veronese maps give similar embeddings. In general, the image of the Veronese mapping $\mathbb{P}^n \xrightarrow{\nu_d} \mathbb{P}^m$ is the projective subvariety of $\mathbb{P}^m$ defined by the polynomials

$$\{z_I z_J - z_K z_L \mid I, J, K, L \in \mathbb{N}^{n+1}, I + J = K + L\},$$

where the z_I are the homogeneous coordinates of $\mathbb{P}^m$ with multi-index notation as described in the proof of the previous proposition. One can prove this by considerations similar to those in the example $\mathbb{P}^1 \xrightarrow{\nu_2} \mathbb{P}^2$.

Example: The image of the Veronese map $\mathbb{P}^1 \xrightarrow{\nu_d} \mathbb{P}^d$ is a curve in $\mathbb{P}^d$ isomorphic to $\mathbb{P}^1$ as a quasi-projective variety. It is called the *rational normal curve of degree d*. The remark above ensures that the defining equations of the rational normal curve in $\mathbb{P}^d$ are the 2×2 subdeterminants of the $2 \times d$ matrix

$$\begin{bmatrix} z_{0,d} & z_{1,d-1} & \cdots & z_{d-1,1} \\ z_{1,d-1} & z_{2,d-2} & \cdots & z_{d,0} \end{bmatrix}.$$

Veronese maps ν_d can be defined for any quasi-projective variety V. One simply considers V as a subset of some projective space $\mathbb{P}^n$ and defines the Veronese map on V to be the restriction of the Veronese map on $\mathbb{P}^n$. The same proof shows that ν_d will define an isomorphism between V and its image.

Because the Veronese mappings define nontrivial (that is, not a mere change of coordinates) isomorphisms of quasi-projective varieties, they are a useful source of examples demonstrating that various properties of projective varieties may not be preserved under isomorphism. We will see an example of this phenomenon in Section 5.5 when we discuss the degree of a projective variety.

Exercise 5.1.1. Consider the Veronese map $\mathbb{P}^2 \xrightarrow{\nu_2} \mathbb{P}^5$. Its image is called the *Veronese surface*. Describe the images of the lines in $\mathbb{P}^2$ on the Veronese surface.

5.2   Five Points Determine a Conic

The following theorem is a simple example of *enumerative algebraic geometry*. This type of algebraic geometry was popular in the nineteenth and early twentieth centuries, especially with the Italian school, and is currently enjoying a renaissance today. A typical goal of enumerative algebraic geometry is to count varieties with certain properties; for example, the number of lines in three-space that intersect four given lines[1] or the number of conics

---

[1]For a detailed answer to this fun problem, see [25].

through four points tangent to a given line. These problems can usually be rephrased so that one is counting the number of points in the intersection of various varieties. One might also wish to count, for example, the number of lines lying on a cubic surface[2]. A more sophisticated question along the same lines would be to count the curves isomorphic to $\mathbb{P}^1$ on a particular surface. "Counting curves" on algebraic varieties is a method for classifying varieties up to isomorphism that is a focus of current research.

The following theorem answers the elementary enumerative question, how many conics (possibly degenerate) pass through five general points in the plane?

Theorem: Given any five points in $\mathbb{P}^2$, there exists a conic containing them all. This conic is unique unless four of the points are collinear, and it is nondegenerate unless three of the points are collinear.

By a *degenerate* conic we mean that the conic is the union of two lines in $\mathbb{P}^2$, or that it is a "double line." Equivalently, the conic is degenerate if its equation factors into linear factors, with the "double line" case being the case where the factors are not distinct.

Proof: A conic in projective space is the zero set of a quadratic homogeneous polynomial:

$$\mathbb{V}(ax^2 + by^2 + cz^2 + dxy + exz + fyz) \subset \mathbb{P}^2.$$

Here the coefficients $a, b, c, d, e, f \in \mathbb{C}$ are not all zero. Multiplying the coefficients a, b, c, d, e, f by some common factor λ produces a different quadratic polynomial, but it defines the same conic in $\mathbb{P}^2$. That is, each line through the origin in $\mathbb{C}^6$, denoted by $[a : b : c : d : e : f]$, defines a conic in $\mathbb{P}^2$. Furthermore, no two distinct lines in $\mathbb{C}^6$ define the same conic. Therefore, we can naturally identify the set of (possibly degenerate) conics in $\mathbb{P}^2$ with projective space $\mathbb{P}^5$. We say that $\mathbb{P}^5$ parametrizes the conics in $\mathbb{P}^2$.

The conic sections passing through a fixed point $[\alpha : \beta : \gamma] \in \mathbb{P}^2$ form a hyperplane H in $\mathbb{P}^5$. This is easily seen by substituting $[x : y : z] = [\alpha : \beta : \gamma]$ into the equation for the conic $ax^2 + by^2 + cz^2 + dxy + exz + fyz = 0$, which leads to a linear equation satisfied by $[a : b : c : d : e : f]$. So the conics passing through P_1, P_2, P_3, P_4, and P_5 form an intersection of hyperplanes $H_1 \cap H_2 \cap H_3 \cap H_4 \cap H_5 \subset \mathbb{P}^5$. With each successive intersection, the dimension drops by one (unless the linear form is dependent on its predecessors, in which case the intersection is unchanged), so the intersection is nonempty. The points in $\mathbb{P}^5$ in this intersection correspond to the conics passing through $P_1, \ldots, P_5$. If the hyperplanes are linearly independent, the intersection $H_1 \cap H_2 \cap H_3 \cap H_4 \cap H_5$ contains exactly one point, so there is

---

[2]The answer, by the way, is twenty-seven; see [20, V.4].

exactly *one* conic passing through the five points. We leave the degenerate case where the hyperplanes are not linearly independent as an exercise.   □

It was precisely these kinds of degenerate special cases that led to problems in some proofs in algebraic geometry around the turn of the century. In the 1930s and 1940s Zariski, Weil and others reworked the foundations of algebraic geometry to put these enumerative results back on a firm footing.

Although interest in enumerative geometry waned in the twentieth century, there has been a burst of activity in this field in the last decade. This is due to the deep connections that have been discovered between enumerative algebraic geometry and theoretical physics (see [18]). Both Edward Witten, a physicist, and Maxim Kontsevich, a mathematician, have been honored with Fields medals because of their work in this area.

Exercise 5.2.1. Given five points in $\mathbb{P}^2$ as above, prove that there is a *unique* conic passing through these points unless four of the points lie on a line. Show that the conic is nondegenerate unless three of the points lie on a line.

Exercise 5.2.2. Show that the set of all nondegenerate conics form a nonempty Zariski-open set of the parameter space $\mathbb{P}^5$ of all conics. Furthermore, show that the set of double lines forms a Zariski-closed set of $\mathbb{P}^5$ isomorphic to the *Veronese surface* (the image of the Veronese map $\mathbb{P}^2 \xrightarrow{\nu_2} \mathbb{P}^5$).

Exercise 5.2.3. Given four points and a line in $\mathbb{P}^2$, show that typically two conics pass through the four points and are tangent to the line. Under what special conditions on the positions of the points and the line do we fail to get exactly two? (Hint: As we will discuss in detail in Section 6.1, a line is tangent to a conic if the defining quadratic function has a double root when restricted to the line; on the other hand, a quadratic polynomial has a double root if and only if its discriminant, a degree two polynomial in its coefficients, is zero.)

Exercise 5.2.4. How many conics in $\mathbb{P}^2$ do we expect through three given points and tangent to two given lines? Through two points and tangent to three lines? Tangent to five lines?

5.3   The Segre Map and Products of Varieties

The Segre map is an important tool that enables us to define, in a natural way, the structure of a quasi-projective variety on a Cartesian product $V \times W$ of two quasi-projective varieties. It is important to remember that the naive approach does not work: The product topology induced on $\mathbb{A}^1 \times \mathbb{A}^1$ is *not* the Zariski topology on $\mathbb{A}^2$ (see Exercise 1.2.2). The Segre map will embed the set $\mathbb{P}^n \times \mathbb{P}^m$ as a closed subset of a higher-dimensional projective space in a natural way, enabling us to speak of the product $\mathbb{P}^n \times \mathbb{P}^m$ as a projective variety.

Let us begin with a low-dimensional example.

Example: The *Segre map* $\Sigma_{1,1}$ is the map

$$\mathbb{P}^1 \times \mathbb{P}^1 \xrightarrow{\Sigma_{1,1}} \mathbb{P}^3,$$
$$([s:t],[u:v]) \longmapsto [\underbrace{su}_{x} : \underbrace{sv}_{y} : \underbrace{tu}_{z} : \underbrace{tv}_{w}].$$

This map is well-defined because the coordinates of the image do not all simultaneously vanish, and because it does not depend on the choice of the representatives $[s:t]$ and $[u:v]$ for elements of $\mathbb{P}^1 \times \mathbb{P}^1$.

If the homogeneous coordinates of $\mathbb{P}^3$ are denoted by x, y, z, and w, then one easily checks that the image of the Segre map $\Sigma_{1,1}$ is the quadric surface $\mathbb{V}(xw - yz) \subset \mathbb{P}^3$.

Using local affine coordinates the map $\Sigma_{1,1}$ takes the form

$$([s:1],[u:1]) \longmapsto [su:s:u:1];$$

that is,

$$(s,u) \longmapsto (su,s,u).$$

From this we see that the image is a *ruled surface,* that is, it can be covered by a family of disjoint lines. To see this, first fix s so that the mapping $u \mapsto [su:s:u:1]$ parametrizes a line on the image surface, then vary s to see the family of disjoint lines. Interchanging the roles of s and u presents the quadric surface as the union of a different family of disjoint lines. The quadratic surface $\mathbb{V}(xw - yz)$, depicted in Figure 5.1, can be written as the disjoint union of lines in two different ways, which is exactly what we expect of the product of two lines.

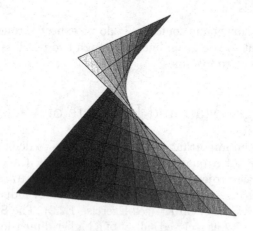

Figure 5.1. The ruled surface $\mathbb{V}(xw - yz)$

Definition: We define the general *Segre mapping* $\Sigma_{m,n}$ by

$$\mathbb{P}^m \times \mathbb{P}^n \xrightarrow{\Sigma_{m,n}} \mathbb{P}^{(m+1)(n+1)-1},$$

$$([x_0 : \cdots : x_m], [y_0 : \cdots : y_n]) \longmapsto [\underbrace{x_0 y_0}_{z_{00}} : \underbrace{x_0 y_1}_{z_{01}} : \cdots : \underbrace{x_i y_j}_{z_{ij}} : \cdots : \underbrace{x_m y_n}_{z_{mn}}].$$

It is easy to remember how this map is defined: Label the homogeneous coordinates of $\mathbb{P}^{(m+1)(n+1)-1}$ as z_{ij}, where $0 \leq i \leq m$ and $0 \leq j \leq n$, and notice that the image of $\Sigma_{m,n}$ is given by

$$
\begin{bmatrix}
z_{00} & \cdots & z_{0n} \\
\vdots & \vdots & \vdots \\
z_{m0} & \cdots & z_{mn}
\end{bmatrix}
=
\begin{bmatrix}
x_0 \\
\vdots \\
x_m
\end{bmatrix}
[y_0 : \cdots : y_n].
$$

Theorem: The image of the Segre map $\Sigma_{m,n}$ is a projective variety defined by the 2×2 minors of the matrix

$$\{(z_{ij})\},$$

where the z_{ij}'s are the homogeneous coordinates of the projective space $\mathbb{P}^{(m+1)\times(n+1)-1}$, doubly indexed and arranged in matrix form. Furthermore, the Segre map is one-to-one, and the projection taking (z_{ij}) to any one of its nonzero columns $[z_{0j} : z_{1j} : \ldots : z_{mj}]$ induces a morphism from the Segre image onto $\mathbb{P}^m$. Likewise, the projection taking (z_{ij}) to any one of its nonzero rows $[z_{i0} : z_{i1} : \ldots : z_{in}]$ induces a morphism from the Segre image onto $\mathbb{P}^n$.

Proof: The image of the Segre map consists of the $(m+1) \times (n+1)$ matrices obtained by multiplying the matrices

$$
\begin{bmatrix}
x_0 \\
\vdots \\
x_m
\end{bmatrix}
\begin{bmatrix}
y_0 & \cdots & y_n
\end{bmatrix}.
$$

In particular, each column of this product matrix is a multiple of every other column, which is to say, the product matrix has rank 1. Of course, the 2×2 subdeterminants of any rank 1 matrix must vanish, so the image of the Segre mapping is contained in the set defined by the 2×2 minors of the matrix (z_{ij}).

On the other hand, suppose a point in $\mathbb{P}^{(m+1)(n+1)-1}$ satisfies the equations $z_{ij} z_{kl} - z_{il} z_{kj} = 0$ for all indices $0 \leq i, k \leq m$ and $0 \leq j, l \leq n$. Arranged into matrix form, the coordinates of this point form a matrix all of whose 2×2 minors vanish. This is equivalent to the condition that the matrix (z_{ij}) has rank at most one. But we know from linear algebra that every $(m+1) \times (n+1)$ matrix of rank k factors as a product of an $(m+1) \times k$ matrix and a $k \times (n+1)$ matrix. In particular,

$$(z_{ij}) = \begin{bmatrix} x_0 \\ \vdots \\ x_m \end{bmatrix} \begin{bmatrix} y_0 & \cdots & y_n \end{bmatrix},$$

for some choice of vectors $[x_0, \ldots, x_m]$ and $[y_0, \ldots, y_n]$, determined up to scalar multiple. Since (z_{ij}) is not the zero matrix, neither (x_i) nor (y_j) is the zero matrix. Thus, (z_{ij}) is in the image of the Segre map.

We now consider the projection from the Segre image onto $\mathbb{P}^m$ and $\mathbb{P}^n$. Again think of the coordinates of $\mathbb{P}^{(m+1)(n+1)-1}$ as the entries in an $(m+1) \times (n+1)$ matrix. Because the image of $\mathbb{P}^m \times \mathbb{P}^n$ consists only of the rank-one matrices, all columns of the matrix (z_{ij}) representing a point in $\mathbb{P}^m \times \mathbb{P}^n$ are multiples of each other. The projection

$$\Sigma(\mathbb{P}^m \times \mathbb{P}^n) \xrightarrow{\pi_1} \mathbb{P}^m,$$
$$(z_{ij}) \longmapsto [z_{0j} : \ldots : z_{mj}],$$

is defined by mapping (z_{ij}) to any of its nonzero columns. Because the columns differ only up to multiplication by a scalar, this map is well defined. The projection $\Sigma(\mathbb{P}^m \times \mathbb{P}^n) \xrightarrow{\pi_2} \mathbb{P}^n$ is defined similarly, with rows instead of columns. $\square$

Given two quasi-projective varieties, $X \subset \mathbb{P}^m$ and $Y \subset \mathbb{P}^n$, the Segre mapping allows us to define the structure of a quasi-projective variety on the product $X \times Y$: We simply restrict the Segre map to the subset $X \times Y$ (see Exercise 5.3.2). When speaking of the Segre product, or simply the product, of two quasi-projective varieties X and Y, we always mean this quasi-projective variety in $\mathbb{P}^{(m+1)(n+1)-1}$. We often denote the Segre image by $X \times Y$. The product projection maps

$$X \times Y \xrightarrow{\pi_1} X$$

and

$$X \times Y \xrightarrow{\pi_2} Y$$

are the restrictions of those defined in the theorem.

Exercise 5.3.1. Fix any point $p = [\lambda_0 : \ldots : \lambda_n] \in \mathbb{P}^n$. Show that the composition

$$\mathbb{P}^m \xrightarrow{\Sigma} \mathbb{P}^{mn+m+n} \xrightarrow{\pi_1} \mathbb{P}^m,$$
$$x \longmapsto \Sigma(x, p) \longmapsto \pi_1(x, p)$$

defines the identity map on $\mathbb{P}^m$.

Exercise 5.3.2. If X and Y are projective varieties, show that the Segre product $X \times Y$ is also a projective variety by expressing its defining equations in terms of those for X and Y. Show that the product of two quasi-projective varieties is quasi-projective.

Exercise 5.3.3. Show that the product of two affine varieties is affine. Note that even in the affine case, our *definition* of product uses the Segre map, the affine varieties being thought of as quasi-projective varieties in some projective spaces.

Exercise 5.3.4. Show that the topology defined on the product above is *not* the product topology, except when one of the varieties is just a finite collection of points.

Exercise 5.3.5. Show that the subset of all degenerate plane conics naturally forms a closed subset in $\mathbb{P}^5$ of dimension four, isomorphic to the projection, from $\mathbb{P}^8$, of the Segree four-fold Σ_{22}, where Σ_{22} denotes the image of $\mathbb{P}^2 \times \mathbb{P}^2$ under the Segre map. (See also Exercise 5.2.2.)

Exercise 5.3.6. Let X and Y be quasi-projective varieties and let $\pi_1 : X \times Y \longrightarrow X$ and $\pi_2 : X \times Y \longrightarrow Y$ be the two natural projections defined in this section. Show that the Segre product $X \times Y$ enjoys the following universal property of products: If Z is any quasi-projective variety admitting morphisms $p_1 : Z \longrightarrow X$ and $p_2 : Z \longrightarrow Y$, then there is a unique map $\mu : Z \longrightarrow X \times Y$ such that the compositions $\pi_i \circ \mu$ agree with p_i.

5.4   Grassmannians

Grassmannians are natural generalizations of projective spaces, and share many of their properties.

Definition: The Grassmannian $\mathbf{Gr}(k, n)$ is the set of all k-dimensional vector subspaces of $\mathbb{C}^n$.

The simplest example of a Grassmannian is the set of one-dimensional subspaces in $\mathbb{C}^{n+1}$, the projective space $\mathbf{Gr}(1, n + 1) = \mathbb{P}^n$.

Grassmannians can be thought of as a set of linear subvarieties of a projective space. A *linear subvariety* of $\mathbb{P}^n$ is a closed subvariety defined by *linear* homogeneous polynomials. An m-dimensional linear subvariety of $\mathbb{P}^n$ is a projective subvariety determined by an $(m + 1)$-dimensional vector subspace of the vector space $\mathbb{C}^{n+1}$. Of course, the projective closure of a linear subvariety in $\mathbb{A}^n$ is a linear subvariety in $\mathbb{P}^n$.

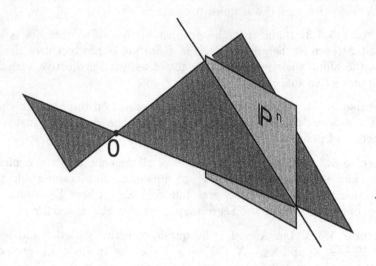

Figure 5.2. A line in $\mathbb{P}^n$ is a 2-dimensional subspace of $\mathbb{C}^{n+1}$

Since a k-dimensional linear subspace of $\mathbb{C}^{n+1}$ is essentially the same as a $(k-1)$-dimensional linear subvariety in $\mathbb{P}^n$, we can think of a Grassmannian $\mathbf{Gr}(k, n)$ as the set of all $(k-1)$-dimensional linear subvarieties of projective space $\mathbb{P}^{n-1}$. This is analogous to thinking of $\mathbb{P}^n$ as the set of points in $\mathbb{P}^n$ or as the set of lines in $\mathbb{C}^{n+1}$. Because of this, some mathematicians use the notation $\mathbf{Gr}(k-1, n-1)$ for our $\mathbf{Gr}(k, n)$.

Grassmannians are themselves projective varieties. This fact follows from the following theorem.

Theorem: The Grassmannian $\mathbf{Gr}(k, n)$ can be embedded as a complex submanifold of $\mathbb{P}^{\binom{n}{k}-1}$.

Proof: Let $\Lambda \in \mathbf{Gr}(k, n)$ be a k-dimensional vector subspace in $\mathbb{C}^n$. Choose basis vectors $(a_{j1}, \ldots, a_{jn})$, $j = 1, \ldots, k$, for Λ, and form the row matrix of basis vectors

$$\begin{bmatrix} a_{11} & \cdots & a_{1n} \\ \vdots & & \vdots \\ a_{k1} & \cdots & a_{kn} \end{bmatrix}.$$

This matrix has full rank, since its rows are linearly independent. Two matrices of full rank (a_{ij}) and (b_{ij}) span the same subspace if and only if there exists a matrix $g \in \mathbf{GL}(k) = \{\text{invertible } k \times k \text{ matrices}\}$ satisfying $(a_{ij}) = g\,(b_{ij})$. Therefore we can identify the set $\mathbf{Gr}(k, n)$ with the factor set

$$G = \{k \times n \text{ matrices of rank } k\}/\text{action of } \mathbf{GL}(k).$$

Let us denote the $k \times k$ subdeterminant of (a_{ij}) formed by the columns $1 \le i_1 < \cdots < i_k \le n$ by $\Delta_{(i_1,\ldots,i_k)}$. The mapping

$$\begin{bmatrix} a_{11} & \cdots & a_{1n} \\ \vdots & & \vdots \\ a_{k1} & \cdots & a_{kn} \end{bmatrix} \longmapsto \left[\Delta_{(1,\ldots,k)} : \cdots : \Delta_{(i_1,\ldots,i_k)} : \cdots : \Delta_{(n-k+1,\ldots,n)} \right]$$

is well-defined on the factor set G. To see this, note that any two equivalent matrices (a_{ij}) and $g(a_{ij})$ are mapped to the same point, as the action of g on the determinants is just multiplication by the nonzero constant $\det g$. Also, because the matrix (a_{ij}) has full rank, at least some determinant is nonzero. So we have a well-defined map $\mathbf{Gr}(k,n) \to \mathbb{P}^{\binom{n}{k}-1}$. Furthermore, it is easy to see that this map is one-to-one. It is known as the *Plücker embedding*.

Under this identification, $\mathbf{Gr}(k,n)$ is at least a subset of projective space. We now want to give it the structure of a complex manifold. That is to say, we want to equip it with an atlas. This will allow us to think of $\mathbf{Gr}(k,n)$ as both a complex manifold and, by Chow's theorem, as a projective variety.

Consider a subspace Λ, where the corresponding matrix (a_{ij}) satisfies $\Delta_{(1,\ldots,k)} \neq 0$. This kind of subspace Λ corresponds to a unique matrix of the form

$$\begin{bmatrix} 1 & \cdots & 0 & a_{1\,k+1} & \cdots & a_{1n} \\ \vdots & \ddots & \vdots & \vdots & & \vdots \\ 0 & \cdots & 1 & a_{k\,k+1} & \cdots & a_{kn} \end{bmatrix},$$

as we can see by multiplying by $g \in \mathbf{GL}(k)$, the inverse of the $k \times k$ matrix formed by the first k columns of the matrix representing Λ. Each matrix of this form determines a unique subspace $\Lambda \in \mathbf{Gr}(k,n)$. So there is a bijective map

$$U_{(1,\ldots,k)} = \{\Lambda \in \mathbf{Gr}(k,n) \mid \Delta_{(1,\ldots,k)} \neq 0\} \longrightarrow \mathbb{C}^{k(n-k)}.$$

Because the open sets $U_{(i_1,\ldots,i_k)}$ where $\Delta_{(i_1,\ldots,i_k)} \neq 0$ cover $\mathbf{Gr}(k,n)$, these mappings form an atlas of $\mathbf{Gr}(k,n)$. We can use the same technique that we used to show that $\mathbb{P}^n$ is a complex manifold to prove that the chart changes are given by multiplications by the rational functions Δ_I/Δ_J, hence the chart changes are manifestly analytic. Also, because the Plücker embedding is given by holomorphic (in fact, rational) maps on the coordinate charts, we know that $\mathbf{Gr}(k,n)$ is described as a complex submanifold of projective space. $\qquad\square$

Theorem: $\mathbf{Gr}(k,n) \subset \mathbb{P}^{\binom{n}{k}-1}$ is a projective algebraic variety.

We have already checked that the Grassmannian $\mathbf{Gr}(k, n)$ is, abstractly, a complex manifold and that it embeds in $\mathbb{P}^{\binom{n}{k}-1}$ as a complex manifold. We can invoke Chow's theorem to conclude that $\mathbf{Gr}(k, n)$ is a projective variety: It is defined by the vanishing of a collection of homogeneous polynomials in $\binom{n}{k}$ variables. Indeed, since the chart transformations are given by regular functions on the algebraic varieties $\mathbb{A}^{k(n-k)}$, we should expect the Grassmannian to be an "abstract algebraic variety" in analogy with a manifold. We do not develop this point of view here, preferring to refer the interested reader to the appendix.

Alternatively, one can easily verify that the image of the Plücker map is a projective variety by finding homogeneous polynomials vanishing precisely on the image (see [17, page 65]). Finding the radical ideal of all polynomials vanishing on $\mathbf{Gr}(k, n) \subseteq \mathbb{P}^{\binom{n}{k}-1}$ is somewhat more involved. Let us say only that the generators for this ideal are certain quadratic polynomials called Plücker relations. They can be derived from certain simple identities about determinants (see [13, page 132]).

Instead of going into these details, we introduce an alternative way of looking at the Grassmannian varieties. This interpretation is based on exterior products.[3] Let $\wedge^k V \subset \wedge^k(\mathbb{C}^n)$ be the kth exterior power of a k-dimensional vector space $V \subset \mathbb{C}^n$. We define the mapping

$$\mathbf{Gr}(k, n) \xrightarrow{\varphi} \mathbb{P}(\wedge^k\mathbb{C}^n) = \mathbb{P}^{\binom{n}{k}-1},$$
$$V \longmapsto \wedge^k V.$$

If the vectors $v_1, \ldots, v_k$ are a basis for $V \subset \mathbb{C}^n$, then $\wedge^k V$ is the one-dimensional subspace of $\wedge^k\mathbb{C}^n$ spanned by $v_1 \wedge \cdots \wedge v_k$. As a mapping into projective space, φ is well-defined, since a subspace V determines its volume form $\varphi(V)$ uniquely up to a constant.

To verify that φ is an isomorphism onto its image set involves some calculations, which we omit (see [17, page 63]).

Exercise 5.4.1. Fix an irreducible conic C in $\mathbb{P}^2$. Show that the set of lines in $\mathbb{P}^2$ that fail to meet the conic in exactly two distinct points is a closed subvariety of the Grassmannian of all lines in $\mathbb{P}^2$, $\mathbf{Gr}(2, 3)$.

5.5   Degree

Classifying all projective varieties up to projective equivalence (up to change of coordinates) is a nearly impossible task. Nonetheless, algebraic geometers have spent the last century trying to sort out this problem. As a first step, it helps to identify invariants of projective varieties: Numbers (or

---

[3]For basic information about exterior algebras, see a linear algebra book, such as [24].

other types of data) that partition the collection of projective subvarieties of $\mathbb{P}^n$. One such numerical invariant is the "degree."

The degree of a variety is defined by intersecting the variety with a linear subvariety of the appropriate dimension and counting the number of intersection points.

Definition: The *degree* of the projective variety V in $\mathbb{P}^n$ is the greatest possible finite number of intersection points of V with a linear subvariety $L \subset \mathbb{P}^n$ of dimension equal to the codimension of V:

$$\deg V = \max\{\#(V \cap L) < \infty \mid L \text{ linear in } \mathbb{P}^n, \ \dim L + \dim V = n\}.$$

In fact, the maximal number of intersection points is almost always achieved: The degree of V is the number of points common to V and a $(\mathrm{codim}V)$-dimensional *generic* linear subvariety. One should interpret the word "generic" here to mean the intuitive idea of a typical, representative, or "sufficiently general" linear subvariety. To make this idea precise, the reader should prove that there is a dense open subset U of the Grassmannian of all $(\mathrm{codim}V)$-dimensional subspaces of $\mathbb{P}^n$ such that for any Λ in this open set, $V \cap \Lambda$ consists of precisely $d = \deg V$ distinct points. In this case, "generic" would mean simply "member of U."

Some examples will clarify the concept of degree.

Example: The degree of the conic $\mathbb{V}(yz - x^2) \subset \mathbb{P}^2$ is two, because a typical line meets a conic in two points. In Figure 5.3, the conic is depicted as a parabola in the affine chart $\mathbb{A}^2$ where $z \neq 0$, and a typical line is shown intersecting the parabola in two points.

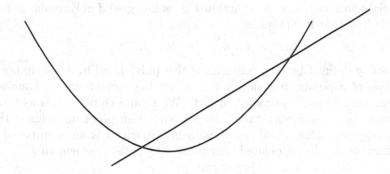

Figure 5.3. Generic line intersecting a conic

Although in a real drawing the line may miss the parabola, in the complex case the line always intersects the conic at least once, but usually twice. A line parallel to the y-axis has one of its intersection points "at infinity," in which case a different choice of affine chart will reveal the two intersection points. The only other possibility is that the line is tangent to the conic, so that there is only one intersection point. In this example we see that the generic number of intersection points coincides with the maximal number of intersection points, which is two.

Our comments about the parabola extend to arbitrary hypersurfaces.

Theorem: If F is an irreducible homogeneous polynomial of degree d, the degree of the hypersurface $\mathbb{V}(F) \subset \mathbb{P}^n$ is d.

The reason F is assumed irreducible in the above theorem is so that F generates the full ideal of polynomials vanishing on the hypersurface $\mathbb{V}(F)$. For example, the polynomials F and F^2 define the same hypersurface, but the degree of F^2 is twice the degree of F. All that is really needed in this theorem is that F has no repeated factors.

Proof: Given an arbitrary line $L \subset \mathbb{P}^n$, the intersection points of V and L can be identified with the zeros of the polynomial function on L obtained by restricting F to the line L. The restriction of F to L produces a degree-d polynomial on $L \cong \mathbb{C}$; which, by the Fundamental Theorem of Algebra, has d roots. For a generic choice of L these roots are distinct, and correspond to the d intersection points of V with L.   □

At this point, we can begin to appreciate the idea of a *scheme*. Schemes allow us to interpret the intersection of a line and a conic, for example, as two points, even when the line is tangent. One simply counts a tangent point with multiplicity two. The intersection of a line L with the hypersurface $\mathbb{V}(F)$ always consists of a collection of exactly d points, provided that these are counted with the appropriate multiplicities.

To elaborate, consider the function $F|_L$ as a degree d polynomial in one variable t, so that it factors as

$$F(t) = (t - a_1)^{m_1} \cdots (t - a_r)^{m_r}.$$

The variety defined by the vanishing of this polynomial in $\mathbb{A}^1$ is simply a collection of r points, but clearly this is not the correct way to consider the geometric object associated to $F(t)$. We should think of this as a set of points $\{a_1, \ldots, a_r\}$ with multiplicities, with each point a_i assigned the multiplicity m_i. This set of points with multiplicities is an example of a *subscheme* of $\mathbb{A}^1$. Its associated coordinate ring is the quotient ring

$$\frac{\mathbb{C}[t]}{(F(t))};$$

the ideal generated by $F(t)$ consists of functions that vanish at a_i to order m_i for all $i = 1, \ldots, r$. This ring differs from the coordinate ring of a variety in that it may have nilpotent elements. The geometric object associated to this coordinate ring is the simplest example of a scheme.

Schemes arise naturally as degenerations of varieties. Here, the variety of d distinct points degenerates into a scheme when some of these points are brought together. This suggests that we may be forced to consider schemes, even if varieties are our primary interest. Just as subvarieties of $\mathbb{A}^n$ correspond to radical ideals in the ring $\mathbb{C}[x_1,\ldots,x_n]$, subschemes of $\mathbb{A}^n$ correspond to arbitrary ideals of $\mathbb{C}[x_1,\ldots,x_n]$.

We now return to our discussion of degree. Degree is important because it helps us partition the subvarieties of $\mathbb{P}^n$ into classes within which we may begin to investigate which are projectively equivalent.

Proposition: The degree of a projective variety is a *projective invariant*: If $\mathbb{P}^n \xrightarrow{T} \mathbb{P}^n$ is an automorphism, then the varieties $V \subset \mathbb{P}^n$ and $T(V) \subset \mathbb{P}^n$ have the same degree.

Proof: An automorphism of $\mathbb{P}^n$ is nothing more than a linear change of coordinates (see Section 3.5), and linear changes of coordinates preserve linear subspaces. □

Some caution is necessary in dealing with degree. Degree is not an invariant of the isomorphism class of a projective variety: Isomorphic subvarieties of $\mathbb{P}^n$ may very well have different degrees. The rational normal curves provide an example of this phenomenon.

Example: Recall that the Veronese mapping

$$\mathbb{P}^1 \xrightarrow{\nu_d} \nu_d(\mathbb{P}^1) \subset \mathbb{P}^d,$$
$$[s:t] \longmapsto [s^d : s^{d-1}t : \cdots : t^d],$$

is an isomorphism onto its image set. The image is called the rational normal curve of degree d because it has degree d as a subvariety of $\mathbb{P}^d$. The reader should verify this fact by intersecting the curve with a generic *hyperplane* in $\mathbb{P}^d$ (that is, a linear subspace defined by a single homogeneous polynomial $\lambda_0 x_0 + \cdots + \lambda_d x_d = 0$). On the other hand, the *linear embedding*

$$\mathbb{P}^1 \longrightarrow \mathbb{P}^1 \subset \mathbb{P}^d,$$
$$[s:t] \longmapsto [s:t:0:\cdots:0],$$

makes $\mathbb{P}^1$ a degree-1 curve (a line) in $\mathbb{P}^d$, since it intersects a hyperplane in exactly one point (except in the "atypical" case where the image is contained in the hyperplane). Thus, although the rational normal curve in $\mathbb{P}^d$ and the line in $\mathbb{P}^d$ are isomorphic as projective varieties (both being isomorphic to $\mathbb{P}^1$), they have different degrees as subvarieties of $\mathbb{P}^d$ and hence are not projectively equivalent.

We can verify these claims about the degree of the rational normal curves pictorially.

We have already seen in Figure 5.3 that the intersection of a line and a conic is typically a set of two points. We also verified in Section 5.1 that the

rational normal curve of degree two is simply the conic $\mathbb{V}(xz - y^2) \subset \mathbb{P}^2$.
We conclude that the rational normal curve in $\mathbb{P}^2$ indeed has degree two.

The rational normal curve of degree 3 is the image set of the Veronese
mapping

$$\mathbb{P}^1 \xrightarrow{\nu_3} \mathbb{P}^3,$$
$$[s : t] \longmapsto [s^3 : s^2 t : st^2 : t^3].$$

On the first coordinate chart, this map is

$$[s : 1] \longmapsto [s^3 : s^2 : s : 1].$$

So the first coordinate chart is mapped to $V = \{(s^3, s^2, s) \in \mathbb{A}^3 \mid s \in \mathbb{C}\} = \mathbb{V}(z^2 - y, z^3 - x)$, which we recognize as the twisted cubic. As we
discussed in Section 3.3, the twisted cubic is the intersection of the two
surfaces $\mathbb{V}(z^2 - y)$ and $\mathbb{V}(z^3 - x)$, as depicted in Figure 5.4. This figure also
depicts a typical plane intersecting the twisted cubic curve. As we see, a
typical plane intersects the curve in three points, thus the rational normal
curve in $\mathbb{P}^3$ has degree three.

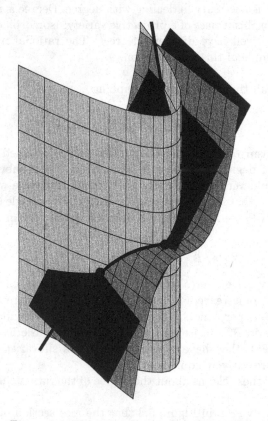

Figure 5.4. The twisted cubic has degree three

Is there any easy way to determine the degree of an arbitrary projective variety? For hypersurfaces this is easy: The degree of the defining polynomial is the degree of the hypersurface. This might lead to a guess that the following could be a general formula:

$$\deg \mathbb{V}(F_1, \ldots, F_c) = \deg F_1 \cdot \deg F_2 \cdots \deg F_c.$$

At least the formula seems to hold for intersections of any number of hyperplanes with a hypersurface of arbitrary degree.

Unfortunately, the situation is not so simple in general. For example, the twisted cubic in $\mathbb{P}^3$ cannot be written as $\mathbb{V}(F_1, \ldots, F_c)$ with $3 = \deg F_1 \cdot \deg F_2 \cdots \deg F_c$. If we could do this, then one of the homogeneous polynomials F_i would have degree 3 and the others would have degree 1. But it is easy to see that no nonzero linear form vanishes on the twisted cubic. Indeed, if $ax + by + cz + dw$ vanishes on the twisted cubic, then $as^3 + bs^2 + cs + d = 0$ for all nonzero $s \in \mathbb{C}$ and so $a = b = c = d = 0$. The only possibility is that the twisted cubic is $\mathbb{V}(F_1)$, a hypersurface in $\mathbb{P}^3$. This is patently false, as a hypersurface in three space has dimension two, while the twisted cubic is a curve, having dimension one.

Although the twisted cubic in $\mathbb{P}^3$ is the intersection of the two surfaces defined by $yw = x^2$ and $z^2w - 2xyz + y^3 = 0$ (see Exercise 3.3.2), the radical homogeneous ideal of V cannot be generated by two elements. However, this ideal can be generated by three polynomials,

$$\mathbb{I}(V) = (x^2 - wy, y^2 - xz, zw - xy) \subseteq \mathbb{C}[x, y, z, w].$$

It can be checked that no two of these polynomials suffice to generate $\mathbb{I}(V)$: Any two cut out a reducible variety consisting of the twisted cubic plus a line.

The problem with the twisted cubic curve in $\mathbb{P}^3$ is that although it has codimension two, its radical homogeneous ideal requires more than two generators. The twisted cubic is not a "complete intersection" of two surfaces.

Definition: A projective variety V in $\mathbb{P}^n$ is called a *complete intersection* if its radical homogeneous ideal $\mathbb{I}(V)$ of all polynomials vanishing on V can be generated by exactly $\mathrm{codim}(V)$ elements.

A complete intersection V is the intersection of $\mathrm{codim}\,V$ hypersurfaces, namely, the hypersurfaces whose equations generate the radical ideal of V. However, the twisted cubic reminds us that not every variety is a complete intersection, even when the variety is the intersection of $\mathrm{codim}(V)$ hypersurfaces. The twisted cubic belongs to the somewhat broader class of varieties that are "set theoretically" like complete intersections.

Definition: A projective variety V of codimension c in $\mathbb{P}^n$ is called a *set-theoretic complete intersection* if V is the intersection of c hypersurfaces.

Of course, every complete intersection is a set-theoretic complete intersection. The difference is that while a set-theoretic complete intersection is defined by an ideal generated by c elements, we do not require that this ideal be radical.

An Open Problem: Much is known about complete intersections, but it remains a difficult problem to identify them. Indeed, we do not even know how to determine whether a given projective variety is isomorphic to a complete intersection. There are a number of interesting questions about set-theoretic complete intersections as well. For example, "Is every irreducible curve in projective three-space the intersection of two surfaces?" Amazingly, this problem remains open despite the best efforts of a number of mathematicians.

In many respects, complete intersections are easier to work with than arbitrary varieties. For example:

Theorem: If $V = \mathbb{V}(F_1, \ldots, F_c)$ is a complete intersection, then

$$\deg V = \deg F_1 \cdot \deg F_2 \cdots \deg F_c,$$

where $c = \operatorname{codim} V$.

The proof of this theorem is not difficult, but it requires developing the rudiments of *intersection theory*. Rather than go into this here, we refer to [37, page 198].

As a special case of this theorem, we consider two curves in the projective plane. This gives us the classical theorem of Bézout.

Bézout's theorem: Consider two curves in $\mathbb{P}^2$ defined by polynomials of degrees d and e, respectively. If they have no common components, their intersection consists of de points. The de points are distinct, provided that the curves are not tangent to each other at any of their intersection points.

In the generic case, the two curves are not tangent at any point, and there are precisely de intersection points. In order to interpret the number of intersection points as exactly de in general, the points must be counted with "multiplicities." The simplest example of Bézout's theorem is given by a generic irreducible curve C (defined by an irreducible polynomial of degree d) intersecting a generic line L (defined by a linear polynomial). We have already seen that $C \cap L$ consists of d distinct points. If the line is in special position relative to C, which in this context means tangent to C at one or more points, the intersection of C with L still consists of d points as long as we count these points with multiplicities. In general, it is somewhat more difficult to assign intersection multiplicities; we return to this subject briefly in Section 6.1. See [14, page 112] or [20, page 54] for a proof of Bézout's theorem.

More generally, algebraic geometers look at intersections of higher-dimensional varieties in $\mathbb{P}^n$ or in some other ambient space. If the varieties have complementary codimension, we "expect" a finite number of intersection points and would like to have a formula or method for computing this *intersection number*. This is the beginning of the beautiful and difficult subject of intersection theory, still an area of active research today.

One context in which it is easy to compute intersection numbers is that of linear varieties in projective space. This suggests the idea of "deforming" a variety in $\mathbb{P}^n$ to a linear subvariety, or more accurately, to a formal combination of linear subvarieties with assigned multiplicities. This is the beautiful and classical topic of "Schubert calculus," accessible in the very readable American Mathematical Monthly article of Kleiman and Laksov [25]. See also [11] for a more advanced, but still elementary, treatment, or [12] for a more thorough treatment of intersection theory.

Exercise 5.5.1. Show that a subvariety of $\mathbb{P}^n$ has degree one if and only if it is a linear subvariety.

Exercise 5.5.2. Find an example of two plane curves that are isomorphic as quasi-projective varieties but that have different degrees. Are the curves projectively equivalent?

Exercise 5.5.3. Find an example of two curves in $\mathbb{P}^2$ that have the same degree but are not isomorphic.

5.6 The Hilbert Function

The degree is a useful invariant for partitioning the set of projective varieties in some fixed $\mathbb{P}^n$ into manageable classes. A related, but much more sophisticated, invariant is the Hilbert polynomial.

Let V be a projective variety in $\mathbb{P}^n$ and let $\mathbb{C}[V]$ be its homogeneous coordinate ring. The collection of all homogeneous polynomials in $\mathbb{C}[V]$ of some fixed degree i forms a finite-dimensional vector subspace R_i of this algebra. The algebra $\mathbb{C}[V]$ is the direct sum of all these subspaces:

$$\mathbb{C}[V] = R_0 \oplus R_1 \oplus R_2 \oplus \cdots ,$$

where $R_0 = \mathbb{C}$ and $R_i R_j \subset R_{i+j}$. In other words, the homogeneous coordinate ring of an algebraic variety is a *graded ring*.

Definition: The *Hilbert function* of the projective variety $V \subset \mathbb{P}^n$ is the function $\mathbb{N} \to \mathbb{N}$ given by $m \mapsto \dim R_m$.

Theorem: For large m the Hilbert function agrees with a polynomial, called the *Hilbert polynomial*,

$$P(m) = e_0 m^d + e_1 m^{d-1} + \cdots + e_d$$

with degree $d = \dim V$ and leading coefficient $e_0 = \frac{\deg V}{d!}$.

For a proof, see almost any book on algebraic geometry or commutative algebra; for example, [20, page 51] or [9, page 43].

Each coefficient of the Hilbert polynomial is a projective invariant of the variety $V \subset \mathbb{P}^n$, meaning that any automorphism of $\mathbb{P}^n$ maps V to a subvariety V' with the same Hilbert polynomial. This gives us a host of new invariants of a projective variety, refining the notion of the degree. In the case of a smooth[4] projective variety $V \subset \mathbb{P}^n$, the Hilbert polynomial is essentially just a compact way to record information provided by the famous *Riemann–Roch formula*. The Riemann–Roch formula tells us how the coefficients of the Hilbert polynomial can be described in terms of intersection numbers (or Chern numbers) for certain vector bundles of differential forms on V and other natural bundles on V. See [20, Appendix].

Although the Hilbert polynomial is unaffected by change of coordinates, it is not an invariant of the isomorphism class of the projective variety—it is not preserved under isomorphism of varieties in general. For instance, the Hilbert function of the variety $\mathbb{P}^1$ is

$$m \longmapsto \dim\left(\mathbb{C}[x,y]\right)_m = m + 1,$$

but the Hilbert function of the twisted cubic $\nu_3(\mathbb{P}^1) \subset \mathbb{P}^3$, which is isomorphic to $\mathbb{P}^1$ as a quasi-projective variety, is

$$m \longmapsto \dim\left(\mathbb{C}[x,y]\right)_{3m} = 3m + 1.$$

The two isomorphic curves $\mathbb{P}^1$ and $\nu_3(\mathbb{P}^1)$ have different Hilbert polynomials, namely $h_1(m) = m+1$ and $h_2(m) = 3m+1$. However, these polynomials have the same degree; this is what we expect, since the degree of the Hilbert polynomial is the dimension of the variety, and dimension is an invariant of the isomorphism class of a variety. The leading coefficients of these polynomials confirm that the degrees of $\mathbb{P}^1$ and the twisted cubic are one and three, respectively, as we have already seen.

Hilbert polynomials are an essential part of the modern theory of classification for algebraic varieties. Indeed, fixing an arbitrary polynomial P, one may ask, what subvarieties of $\mathbb{P}^n$ have Hilbert polynomial P? It turns out that the set of all subvarieties with Hilbert polynomial P form, in a natural way, a quasi-projective variety—or more accurately a scheme—called the *Hilbert scheme*. The Hilbert scheme is thus a parameter space for subvarieties of $\mathbb{P}^n$, and understanding its structure helps us understand the way in which subvarieties of $\mathbb{P}^n$ are related to each other. Interesting questions abound: What is the dimension of this quasi-projective variety? What does it mean if there is a path from one point to another on the Hilbert scheme? What geometric interpretation can be given to the different components

---

[4]We will discuss smoothness in more detail later. For now, we can just think of a smooth variety as a complex manifold.

of the Hilbert scheme? The study of Hilbert schemes is an active area of research in algebraic geometry today.

To get a rough idea of how the Hilbert scheme is constructed, first recall that every variety in $\mathbb{P}^n$ is uniquely defined by its radical homogeneous ideal I in the polynomial ring S in $n+1$ variables. For large r, one can check that I is determined by its elements of degree r. For example, if I is the principal ideal generated by a homogeneous polynomial F, then the degree r elements of I are spanned over k by the elements of the form $x_0^{a_0} x_1^{a_1} \cdots x_n^{a_n} F$ where $\deg F + \sum a_i = r$; from this we can recover F and hence I. Although it is not at all obvious, Grothendieck showed that for a fixed polynomial P, there exists an r that works universally for all ideals I. That is, there exists an r (depending on P) such that for all ideals I defining a variety whose Hilbert polynomial is P, I is the radical of the subideal generated by its elements of degree r. Thus to give a variety with Hilbert polynomial P is to give a vector subspace I_r of the $\binom{n+r}{r}$-dimensional vector space S_r of all homogeneous polynomials of degree r. All these vector subspaces I_r have the same dimension, namely $\dim S_r - \dim(S/I)_r = \binom{n+r}{r} - P(r)$; call this dimension d_r. Thus to give a variety with Hilbert polynomial P is to give a point in the Grassmannian of d_r-dimensional subspaces in the $\binom{r+n}{r}$-dimensional vector space S_r. In this way, every variety in $\mathbb{P}^n$ with Hilbert polynomial P corresponds to a unique point in this Grassmannian. On the other hand, not every point in this Grassmannian corresponds to the ideal of a variety (really, a scheme) with Hilbert polynomial P: It turns out that those that do lie in a closed subvariety of this Grassmannian defined by certain determinantal equations. To carry out the details of this construction is a somewhat technical but worthwhile process; see [19].

It is natural to try to find a parameter space for projective varieties up to projective equivalence. Hilbert schemes do not serve this purpose: Two distinct but projectively equivalent varieties determine two distinct points of the Hilbert scheme. However, the automorphism group $\mathbf{PGL}(n+1)$ of $\mathbb{P}^n$ acts on the set of varieties in $\mathbb{P}^n$ with a fixed Hilbert polynomial, inducing a natural action on each Hilbert scheme. Therefore, the quotient of the Hilbert scheme by this $\mathbf{PGL}(n+1)$ action ought to give us a parameter space for projective varieties up to projective equivalence, at least as we range over all possible Hilbert polynomials. Unfortunately, it is not a simple matter to define the structure of an algebraic variety on this quotient set in general. This leads to the difficult and beautiful subject of *geometric invariant theory*, developed by Fields medalist David Mumford in his quest for moduli spaces for algebraic varieties [16].

Exercise 5.6.1. Assume that the variety $V \subset \mathbb{P}^n$ has the Hilbert polynomial $P(n)$. Calculate the Hilbert polynomial of the image variety $\nu_d(V) \subset \mathbb{P}^{\binom{n+d}{d}-1}$ of the Veronese map.

Exercise 5.6.2. Find the Hilbert polynomial P of a k-dimensional linear subvariety of $\mathbb{P}^n$. Describe the Hilbert scheme of varieties in $\mathbb{P}^n$ with Hilbert polynomial P.

Exercise 5.6.3. Find the Hilbert polynomial of a degree-d hypersurface in $\mathbb{P}^n$. What is the Hilbert scheme of varieties with this Hilbert polynomial?

6

Smoothness

6.1   The Tangent Space at a Point

The tangent space to an algebraic variety at a point can be defined purely algebraically in such a way that it agrees with the concept familiar to students of calculus. The definition generalizes the observation that the tangency of the x-axis to the parabola defined by $y = x^2$ can be detected by the fact that the polynomial function $f(x) = x^2$ has a "double root" at $x = 0$.

Because tangency is a local issue, we discuss tangent spaces first only in the affine case, and assume that our variety V is a Zariski-closed set in some fixed $\mathbb{A}^n$. Furthermore, in considering the tangent space to V at a point p, we first choose our coordinate system so that the point p is the origin.

We begin by considering an arbitrary line ℓ in $\mathbb{A}^n$ passing through the origin and through some fixed point, $q = (a_1, \ldots, a_n)$. This line can be parametrized by the formula $\ell = \{(ta_1, \ldots, ta_n) \mid t \in \mathbb{C}\}$. When is this line tangent to V at the origin?

If $F_1, \ldots, F_r$ are generators for the radical ideal $\mathbb{I}(V)$ defining $V \subset \mathbb{A}^n$, then the intersection of V with ℓ is found by solving for t in the system of equations

$$F_1(ta_1, \ldots, ta_n) = 0,$$
$$\vdots$$
$$F_r(ta_1, \ldots, ta_n) = 0.$$

This intersection will consist of (possibly) several points along ℓ, but because we have assumed that the origin lies on both V and ℓ, we know that at least $t = 0$ is a solution.

Because the point $q = (a_1, \ldots, a_n)$ is fixed, each of the expressions $F_i(ta_1, \ldots, ta_n)$ is a polynomial in the one variable t, and hence factors completely into linear factors. The intersection points $V \cap \ell$ correspond to the common roots. Some of the intersection points may appear "with multiplicities," corresponding to simultaneous multiple roots of the r polynomials $F_i(ta_1, \ldots, ta_n)$ in t. In particular, the *multiplicity* of $V \cap \ell$ at the origin is the exponent of the highest power of t that divides all the polynomials $f_i(t) = F_i(ta_1, \ldots, ta_n)$. This leads to the following definitions.

Definition: The line ℓ *is tangent to V at p* if the multiplicity of $\ell \cap V$ at p exceeds one. Moreover, we say "ℓ *is tangent to V of order n*" if the multiplicity is $n + 1$. The *tangent space $T_p V$ of V at p* is the union of all points lying on lines tangent to V at p. In the degenerate case where p is an isolated point of the variety V, the tangent space $T_p V$ is defined to be the zero-dimensional vector space consisting only of the point p.

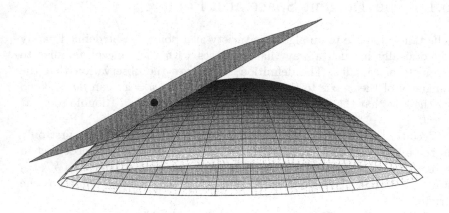

Figure 6.1. Tangent space to an algebraic variety

In order to be sure that these definitions make sense, we need to verify that they are independent of the choice of generators F_i for $\mathbb{I}(V)$. We also want to ensure that the union of all points lying on tangent lines really forms a linear variety in $\mathbb{A}^n$.

First, some examples may help clarify the definition.

Examples: (1) The variety $V \subseteq \mathbb{A}^2$ defined by the equation $y = x^2$ intersects the line ℓ parametrized by $\{(ta, tb) | t \in \mathbb{A}^1\}$ in the subset of ℓ defined by the solutions to

$$tb - t^2a^2 = 0.$$

That is, the intersection consists of 2 points, the origin (corresponding to the solution $t = 0$) and the point $\left(\frac{b}{a}, \left(\frac{b}{a}\right)^2\right)$. However, if $b = 0$, then these points coincide, and ℓ intersects V at the origin with multiplicity 2. Thus, the only tangent line to V at the origin is the line through the origin and a point of the form $(a, 0)$. As expected, the tangent space to the parabola at the origin is the x–axis.

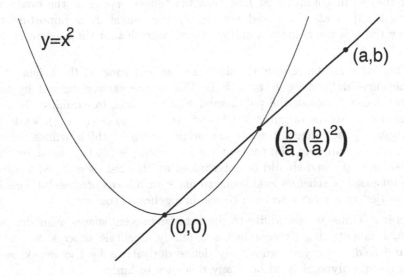

Figure 6.2. The line is tangent only when $b = 0$

(2) The nodal curve $V = \mathbb{V}(y^2 - x^2 - x^3) \subset \mathbb{A}^2$ intersects itself at the origin. The line through the origin and (a, b) is tangent to V at the origin if and only if $t = 0$ is a multiple root of the polynomial $(tb)^2 - (ta)^2 - (ta)^3$. Since for any values of a and b, $t = 0$ is a multiple root, we see that all lines through the origin are tangent to V.

(3) The variety $V = \mathbb{V}(y^2 - x^3)$ also has tangent space at the origin equal to the full affine plane, as the reader will easily check. This is again due to a "singular" point at the origin.

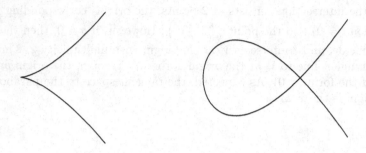

Figure 6.3. Tangent space at a cusp or a node is the plane

It may seem confusing at first that the tangent space at the origin in example (3) is all of $\mathbb{A}^2$ and not simply the x-axis. It is important to follow through the definition and convince yourself that the tangent space is indeed $\mathbb{A}^2$.

There is a related concept, called the *tangent cone at the origin* of a plane curve defined by $F(x,y) = 0$. This is the variety defined by the lowest-degree terms of the polynomial $F(x,y)$. Thus, in example (2), the tangent cone to the origin is $\mathbb{V}(x^2 - y^2) = \mathbb{V}(x - y) \cup \mathbb{V}(x + y)$, while in example (3) the tangent cone to the origin is $\mathbb{V}(y^2)$, which defines the x-axis. More accurately, the tangent cone in example (3) is defined by the vanishing of y^2, and should be interpreted as "the x-axis counted twice." The language of schemes enables us to make such ideas precise, but this is the subject of a more advanced course in algebraic geometry.[1]

As in calculus, we would like to characterize tangent spaces using derivatives. Differentiating polynomials is a purely algebraic process over any ground field, since you can simply define derivatives by the well-known formula for polynomials without any reference to limits.

The *differential* of a polynomial $F \in \mathbb{C}[x_1, \ldots, x_n]$ *at the origin* is its linear part, that is, the sum of the homogeneous terms of degree one in F. The differential at any point p can be defined by choosing coordinates so that the point p is the origin. This is made precise in the following definition.

Definition: The *differential* $dF|_p$ of a polynomial F at an arbitrary point $p = (p_1, \ldots, p_n)$ in $\mathbb{C}^n$ is the linear part of the Taylor series expansion of

---

[1]For example, see [37, pages 79 and 80].

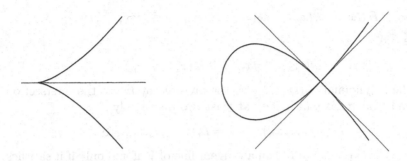

Figure 6.4. The tangent cone to the cusp and to the node

F at p. That is, writing F uniquely in the form

$$F(x) = F(p) + L(x_1 - p_1, \ldots, x_n - p_n) + G(x_1 - p_1, \ldots, x_n - p_n),$$

where L is linear and G is a polynomial with no linear or constant terms, the *differential of F at p* is the linear part $L(x - p)$. The coefficient of the linear term $(x_j - p_j)$ is the partial derivative of F with respect to x_j, evaluated at p. Symbolically, we have

$$L(x - p) = dF|_p(x - p) = \sum_{j=1}^{n} \frac{\partial F}{\partial x_j}(p)\,(x_j - p_j).$$

Recall from calculus that when $F(p) = 0$, $L(x - p)$ is the linear function that best approximates $F(x)$ in a neighborhood of p.

Using differentials, we can explicitly describe the equations of the tangent space to a variety at a point.

Theorem: Let V be an affine algebraic variety in $\mathbb{A}^n$ defined by the vanishing of the polynomials with $F_1, \ldots, F_r$, which we assume to be generators of the radical ideal of V. Let p be a point in V. Then the tangent space to V at p is the linear variety

$$T_p V = \mathbb{V}(dF_1|_p(x - p), \ldots, dF_r|_p(x - p)) \subset \mathbb{A}^n.$$

Moreover, the tangent space is independent of the choice of the generators F_i.

Because it is a linear space, the tangent space $T_p V$ can and should be interpreted as a vector space with the origin at p.

Proof: By choosing appropriate coordinates, p may be taken to be the origin. Consider a line ℓ through the origin and a fixed point $(x_1, \ldots, x_n)$, so that ℓ is parametrized by $\{(tx_1, \ldots, tx_n) \mid t \in \mathbb{C}\}$. Because p (the origin) lies on V, we have $F_i(0) = 0$, so that

$$F_i(tx_1, \ldots, tx_n) = L_i(tx_1, \ldots, tx_n) + G_i(tx_1, \ldots, tx_n),$$

where

$$L_i(tx_1, \ldots, tx_n) = t\, L_i(x_1, \ldots, x_n),$$

and the polynomial $G_i(tx_1, \ldots, tx_n)$ is divisible by t^2. So the intersection of ℓ and V at p has multiplicity at least two if and only if

$$L_1(x_1, \ldots, x_n) = \cdots = L_r(x_1, \ldots, x_n) = 0.$$

We see that $(x_1, \ldots, x_n)$ lies on a tangent line of V if and only if it satisfies these linear equalities. Since the tangent space to V at p is the union of all the points lying on all the tangent lines, we see that the tangent space is precisely the linear variety $\mathbb{V}(L_1, \ldots, L_r) \subseteq \mathbb{A}^n$.

We must verify independence of the choice of generators. So let $\widetilde{F}_1, \ldots, \widetilde{F}_s$ be another set of generators for $\mathbb{I}(V)$. Then

$$F_i = H_{i1}\widetilde{F}_1 + \cdots + H_{is}\widetilde{F}_s$$

for some functions H_{ij} in $\mathbb{C}[x_1, \ldots, x_n]$, so

$$dF_i = (dH_{i1})\widetilde{F}_1 + \cdots + (dH_{is})\widetilde{F}_s + (H_{i1})d\widetilde{F}_1 + \cdots + (H_{is})d\widetilde{F}_s.$$

In particular, since the $\widetilde{F}_i$ vanish at $p \in V$, we have

$$dF_i|_p = H_{i1}(p)d\widetilde{F}_1|_p + \cdots + H_{is}(p)d\widetilde{F}_s|_p.$$

So $\mathbb{V}(dF_1|_p, \ldots, dF_r|_p) \supset \mathbb{V}(d\widetilde{F}_1|_p, \ldots, d\widetilde{F}_r|_p)$, with the reverse inclusion following by symmetry.

Finally, because all the defining equations $dF_i|_p$ are linear, the variety T_pV is a linear subvariety of $\mathbb{A}^n$. $\square$

In order to define the tangent space to a quasi-projective variety V at a point p using this idea, we must think of V as an open subset in a closed subvariety W of some fixed projective space. We can then choose an affine chart $\mathbb{A}^n$ containing p and look at the tangent space to the affine variety $W \cap \mathbb{A}^n$. This will be a linear subspace in the affine chart $\mathbb{A}^n$. This definition is somewhat unsatisfactory, however, because a different choice of affine chart will of course produce a different linear space. On the other hand, any two such choices will have the same closure in $\mathbb{P}^n$. Thus it makes more sense in this case to consider the *projective tangent space* to V at p, which is defined as the projective closure of any of the tangent spaces to V at p in an affine chart. The projective tangent space to V at p is independent of the choice of affine chart.

An alternative way to define the projective tangent space to a quasi-projective variety is as follows. First, because every quasi-projective variety is an open subset of closed subset in $\mathbb{P}^n$, it is enough to define the projective tangent space at a point p on a projective variety V in $\mathbb{P}^n$. Let $\tilde{V}$ be the affine cone over V in $\mathbb{A}^{n+1}$, and let $\tilde{p}$ be any point of $\tilde{V}$ representing the point p on V. Consider the tangent space $T_{\tilde{p}}\tilde{V}$ in $\mathbb{A}^{n+1}$. Because the line through the origin and through $\tilde{p}$ lies in $\tilde{V}$, this line also lies on $T_{\tilde{p}}\tilde{V}$. This means that the linear subvariety $T_{\tilde{p}}\tilde{V}$ of $\mathbb{A}^{n+1}$ passes through the origin of $\mathbb{A}^{n+1}$ and thus gives rise to a unique linear projective subvariety (of dimension one less) in the projective space $\mathbb{P}^n$. Furthermore, the reader can easily check that all points $\tilde{p}$ lying on the same line through the origin in $\mathbb{A}^{n+1}$ share the same tangent space to $\tilde{V}$, so they give rise to the same linear projective variety in $\mathbb{P}^n$. This linear projective variety is the projective tangent space to V at p.

The definition of a projective tangent space is still unsatisfactory, however, because as stated there is no good way to compare the resulting spaces if we interpret our variety V as embedded in different projective spaces. This difficulty can be surmounted by taking a more algebraic and abstract point of view, but we will not pursue this here (see [37, Chapter II, Section 1.4]).

It is clear from our discussion that at least the dimension of the tangent space T_pV makes sense for any quasi-projective variety, and that it is independent of the manner in which we interpret V as a subset of projective space. This dimension is always greater than or equal to the dimension of V at p. Admittedly, this is not obvious, since V may be defined by many more equations than its codimension, and then this is also true of T_pV. For a proof, see Shafarevich [37, Chapter II, Section 1.4].

Exercise 6.1.1. Using the theorem describing the defining equations for T_pV in terms of the equations for V, compute the tangent spaces of the curves in examples (1), (2), and (3) at the origin.

Exercise 6.1.2. Show that the two different ways of defining the projective tangent space to a projective variety in $\mathbb{P}^n$ yield the same space.

Exercise 6.1.3. Let V be a projective variety in $\mathbb{P}^n$ whose homogeneous radical ideal is generated by homogeneous polynomials $F_1, \ldots, F_m$. Show that the projective tangent space to V at p is defined by the homogeneous linear polynomials $dF_1|_p, \ldots, dF_m|_p$. (Hint: Consider the affine cone over V and use the corresponding theorem for affine varieties.)

Exercise 6.1.4. Let $V \subset \mathbb{P}^n$ be a hypersurface defined by a homogeneous irreducible polynomial F. Find an explicit description of the tangent space to V at a point p. What conditions on p ensure that the tangent space to V at p has dimension $n - 1$?

6.2 Smooth Points

We can finally define what it means for a variety to be *smooth*, an important concept the reader probably already has a feeling for. The idea is that a variety is smooth at a point p if the tangent space at p has the expected dimension.

Definition: A point p on a quasi-projective variety V is *smooth* if

$$\dim T_p V = \dim_p V.$$

Otherwise, $p \in V$ is *singular*.

Because the dimension of the tangent space to a variety V at p is independent of the choice of embedding of V in projective space and independent of the affine neighborhood of p used to find the tangent space, the notion of a smooth point of V is intrinsic to V, that is, smoothness is invariant under isomorphism and does not depend on extrinsic features such as a particular embedding as a locally closed subset of $\mathbb{P}^n$.

It is interesting to note that the definition of smoothness is purely algebraic—it makes sense for varieties defined over an arbitrary field.

Examples: The projective space $\mathbb{P}^n$ is smooth at every point, since $\mathbb{P}^n$ has a cover by affine open sets $\mathbb{A}^n$. The tangent space at any point of $\mathbb{A}^n$ is all of $\mathbb{A}^n$, so $\dim T_p \mathbb{P}^n = n = \dim \mathbb{P}^n$ for all points p of $\mathbb{P}^n$. Likewise, Veronese surfaces and rational normal curves (Section 5.1) are smooth at every point, because they are isomorphic as algebraic varieties to $\mathbb{P}^2$ and $\mathbb{P}^1$, respectively, and smoothness is invariant under isomorphism. Similarly, the product variety $\mathbb{P}^n \times \mathbb{P}^m$ is smooth, because any point has a neighborhood isomorphic to $\mathbb{A}^n \times \mathbb{A}^m \cong \mathbb{A}^{n+m}$, which is smooth.

The set of singular points of a variety is easy to describe explicitly. This *singular locus* is a proper closed subset of the variety. Thus, the *smooth locus,* namely, the set of smooth points of an algebraic variety V, is a nonempty Zariski-open subset of V, and so is a very large subset of V.

Theorem: The locus of singular points of a quasi-projective variety V forms a proper closed subset of V. Explicitly, if V is an irreducible affine variety in $\mathbb{A}^n$ of dimension d whose radical ideal $\mathbb{I}(V)$ is generated by $F_1, \ldots, F_r$, then the singular locus of V is the common zero set in V of the polynomials obtained as the $(n-d) \times (n-d)$ minors of the Jacobian matrix

$$\begin{bmatrix} \frac{\partial F_1}{\partial x_1} & \cdots & \frac{\partial F_1}{\partial x_n} \\ \vdots & & \vdots \\ \frac{\partial F_r}{\partial x_1} & \cdots & \frac{\partial F_r}{\partial x_n} \end{bmatrix}.$$

Sketch of Proof: Because a quasi-projective variety has a basis of affine open subvarieties, it suffices to prove that the singular locus of an affine variety is a proper closed subset. Let p be a point on an affine variety V in $\mathbb{A}^n$. The tangent space at $p = (p_1, \ldots, p_n)$ is defined as the zero set of the r linear polynomials $dF_1|_p(x-p), \ldots, dF_r|_p(x-p)$, which can be obtained as the matrix product

$$\begin{bmatrix} \frac{\partial F_1}{\partial x_1}\big|_p & \cdots & \frac{\partial F_1}{\partial x_n}\big|_p \\ \vdots & & \vdots \\ \frac{\partial F_r}{\partial x_1}\big|_p & \cdots & \frac{\partial F_r}{\partial x_n}\big|_p \end{bmatrix} \begin{bmatrix} x_1 - p_1 \\ \vdots \\ x_n - p_n \end{bmatrix}.$$

Thus, we can think of the tangent space as the kernel of a linear map given by the Jacobian matrix. Now, $p \in V$ is singular if and only if the dimension of the tangent space exceeds d. This occurs if and only if the rank of the Jacobian matrix is strictly smaller than $n - d$ at p. The rank is less than $n - d$ if and only if all $(n-d) \times (n-d)$ minors vanish. Thus the singular locus is defined by the vanishing of the minors of the Jacobian matrix as claimed. This shows that the singular locus is a *closed* subvariety of V.

It remains only to show that the singular locus is a *proper* subvariety of V, that is, it is not possible that every point of V is a singular point. We first consider the case where V is a hypersurface, that is, where $V = \mathbb{V}(F) \subseteq \mathbb{A}^n$ for a single polynomial F in n variables. In this case, $\text{Sing}\,V = \mathbb{V}(\frac{\partial F}{\partial x_1}, \ldots, \frac{\partial F}{\partial x_n}) \cap V$. If V is everywhere singular, then each $\frac{\partial F}{\partial x_i}$ must vanish everywhere on V. This means that $\frac{\partial F}{\partial x_i}$ is in the ideal $\mathbb{I}(V) = (F)$ defining V. But since the degree of $\frac{\partial F}{\partial x_i}$ is strictly less than the degree of F, this is impossible (provided that x_i appears in F, but we know that some x_i does).

The proof can be completed by showing that every affine irreducible variety has a dense open subset that is isomorphic to a hypersurface. This is accomplished by a succession of projections of the variety to lower-dimensional spaces. This is not difficult, but we postpone the details until Section 7.5. □

The theorem does not require that V be irreducible. The statement holds whenever V is equidimensional, that is, when all irreducible components have the same dimension.

The theorem implies that an affine variety V is "almost everywhere smooth" or "generically smooth." Since every quasi-projective variety contains a dense open affine subvariety, it is also true that every quasi-projective variety is generically smooth.

In the proof of the theorem, we have, for the first time, made use of the assumption that the characteristic of the field of complex numbers is zero. Indeed, the claim that $\frac{\partial F}{\partial x_i}$ cannot be in the ideal generated by F when x_i appears in F is false in characteristic $p > 0$. For example, if

$F(x, y, z) = x^p + y z$, then $\frac{\partial F}{\partial x} = p\, x^{p-1} = 0$ is in the ideal generated by F. Nonetheless, the theorem remains true over an arbitrary algebraically closed ground field, although some more technical algebra is required for the proof in nonzero characteristic. [2]

It is also possible to describe the singular points of a projective variety, and hence of a quasi-projective variety, in terms of its homogeneous defining equations. Consider an equidimensional projective variety $V \subseteq \mathbb{P}^n$, defined by homogeneous polynomials $F_1, \ldots, F_r \in \mathbb{C}[x_0, \ldots, x_n]$, generating a radical ideal. The reader should be able to show that the previous theorem implies

$$\operatorname{Sing}(V) = \mathbb{V}\left(\text{the } c \times c \text{ subdeterminants of } \left[\frac{\partial F_i}{\partial x_j}\right]\right) \cap V \subset \mathbb{P}^n,$$

where $c = \operatorname{codim} V$ in $\mathbb{P}^n$. Thus, a projective variety is smooth if and only if the corresponding cone-shaped affine variety in one higher dimension has at worst one isolated singularity—the vertex of the cone at the origin in $\mathbb{A}^{n+1}$.

Example: Consider the conic curve in $\mathbb{P}^2$ defined by the homogeneous polynomial $x^2 + y^2 - z^2$. Both this projective plane curve and the affine cone over it have codimension one in their respective ambient spaces, since each is defined by the vanishing of the single polynomial $x^2 + y^2 - z^2$ in $\mathbb{P}^2$ and in $\mathbb{A}^3$, respectively. Thus the singular locus of both the projective curve and the affine cone over it is given by the vanishing of the 1×1 minors of the Jacobian matrix

$$\begin{bmatrix} \frac{\partial F}{\partial x} \\[2mm] \frac{\partial F}{\partial y} \\[2mm] \frac{\partial F}{\partial z} \end{bmatrix} = \begin{bmatrix} 2x \\[2mm] 2y \\[2mm] -2z \end{bmatrix}.$$

Therefore, the singular locus of the projective conic is the subvariety $\mathbb{V}(2x, 2y, -2z)$ in $\mathbb{P}^2$, that is, the empty set. Likewise, the singular locus of the affine cone is the subvariety $\mathbb{V}(2x, 2y, -2z)$ in $\mathbb{A}^3$, the origin. In particular, the affine cone over a smooth projective variety can have a singular point at the vertex.

---

[2]See [37, Chapter II, Section 1.4]

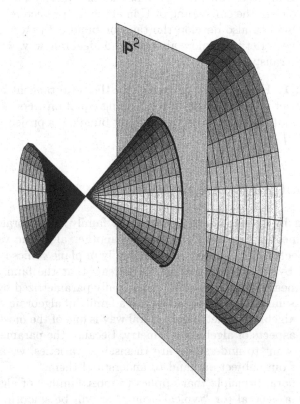

Figure 6.5. A singular cone over a smooth projective variety

Just as in other branches of geometry, it is possible to define a total *tangent bundle* to any smooth affine variety $V \subseteq \mathbb{A}^n$ of dimension d. For each $p \in V$, let $T_pV \subseteq \mathbb{A}^n$ be the tangent space at p. Now consider the set

$$TV = \{(p, y) \mid y \in T_pV\} \subseteq V \times \mathbb{A}^n \subseteq \mathbb{A}^n \times \mathbb{A}^n.$$

We leave it as an easy exercise to check that TV is a closed subvariety of $V \times \mathbb{A}^n$. Furthermore, the natural projection $TV \to V$ produces the structure of a rank-d vector bundle on TV: The fiber over each point $p \in V$ is the variety T_pV. The variety T_pV can be identified with a d-dimensional affine space, together with a marked point p, which we can think of as a d-dimensional vector space (p corresponds to the origin). The variety TV is called the tangent bundle to V. Vector bundles will be discussed in greater detail in the final chapter.

If $V \subseteq \mathbb{P}^n$ is a smooth projective variety, one can similarly construct the total tangent bundle as a *projective space bundle* over V. This is the subvariety of $V \times \mathbb{P}^n$ consisting of pairs (p, T_pV), where now T_pV denotes the projective tangent space to V at p.

It is not obvious from this point of view that the variety TV is independent of the choice of the embedding of V in affine (or projective) space, but this is true. One can also develop the tangent bundle TV to an arbitrary quasi-projective variety in a more abstract and algebraic way, which makes such concerns transparent.

Exercise 6.2.1. Find defining equations for the total tangent bundle of a Zariski-closed set in $\mathbb{A}^n$ in terms of its defining equations. For a projective variety, show that the total projective tangent bundle is a projective variety.

6.3 Smoothness in Families

We have seen in numerous examples that a family of algebraic varieties is often parametrized in a natural way by another algebraic variety. For example, in Section 5.2, we saw that the family of plane conics is naturally parametrized by $\mathbb{P}^5$, and in Section 5.4 we saw that the family of all d-dimensional linear subvarieties of $\mathbb{P}^n$ is naturally parametrized by a variety called the Grassmannian. This tendency of a family of algebraic varieties to itself form an algebraic variety in a natural way is one of the most beautiful and powerful aspects of algebraic geometry. Because the parameter spaces of objects we want to understand are themselves varieties, we again have all the tools of our subject on hand to understand them.

A useful general principle that applies to most families of algebraic varieties is that a general (or "typical") member will be smooth. Indeed, if a family of varieties is parametrized by an irreducible variety V, then we can infer the smoothness of the general member of V from the smoothness of just one member. The reason is that the smooth members form an *open subset* of the parameter space. Like all open sets of an irreducible variety, this subset of smooth members is *dense*, provided that it is nonempty.

We will not prove this general principle, called the "property of generic smoothness." The interested reader can consult [20, Chapter III, Section 10] for more details and a proof. Instead, we illustrate the principle with an example.

Example: Let us look at the one-parameter family of hyperbolas $\{\mathbb{V}(xy - t) \mid t \in \mathbb{C}\}$ in affine space $\mathbb{A}^2$. This family of hyperbolas is naturally parameterized by the points of the variety $\mathbb{A}^1$. We can understand this parameterization as a map π from a variety V to $\mathbb{A}^1$, where the members of the family are the fibers of the map π. More specifically, let V be the affine subvariety of $\mathbb{A}^3$ defined by the vanishing of the polynomial $xy - z$, and let π be the (restriction to V of) the natural projection to the z-coordinate. The parameterization $t \mapsto \mathbb{V}(xy - t)$ assigns to each fixed t in $\mathbb{A}^1$ the fiber $\pi^{-1}(t)$, which is the hyperbola $\mathbb{V}(xy - t)$ in the plane $\mathbb{A}^2 \cong \mathbb{V}(z - t) \subset \mathbb{A}^3$.

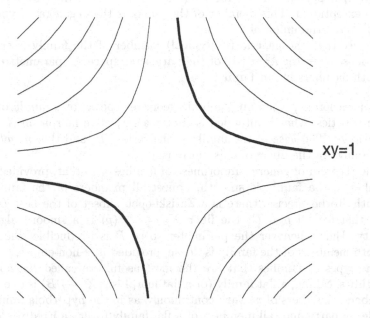

Figure 6.6. A one-parameter family of hyperbolas

The members of our family thus fit together nicely into the algebraic variety V, the surjective morphism π describes the parameterization.

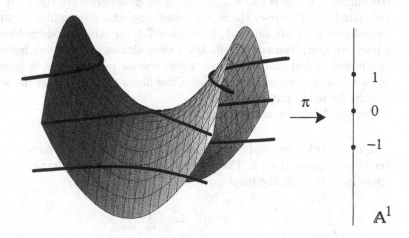

Figure 6.7. The hyperbola family forms a variety $\mathbb{V}(xy - z)$

Note that the fiber over any nonzero point of $\mathbb{A}^1$ is a nondegenerate (smooth) hyperbola of the form $\mathbb{V}(xy - t)$. Only the fiber over the origin is exceptional: This member of the family is the degenerate hyperbola $\mathbb{V}(xy) \subset \mathbb{A}^2$, the union of 2 lines.

We see that the generic (or typical) member of the family is smooth. The dense open set $\mathbb{A}^1 \smallsetminus \{0\}$ of the parameter space $\mathbb{A}^1$ parametrizes the smooth members in our family.

More generally, when an algebraic geometer speaks of a *family* of algebraic varieties, what is intended is simply a surjective morphism $X \xrightarrow{\pi} B$ of varieties. The *base* of the family is the variety B, and the *members* of the family are the fibers of this morphism.

The theorem of generic smoothness of families says that, provided some member of the family is smooth, almost all members of the family are smooth. To be precise, there is a Zariski-open subset of the base $U \subset B$ such that for all $p \in U$, the fiber $X_p = \pi^{-1}(p)$ is a smooth algebraic variety. Thus whenever the parameter space B is irreducible, the set of smooth members of the family is dense, provided it is non-empty.

The types of families that are the most useful are called *flat* families. Roughly speaking, a flat family (or a flat morphism $X \xrightarrow{\pi} B$) is one whose members (the fibers of π) vary continuouly, as in the hyperbola family example. In particular, all members of a flat family (over an irreducible base B) have the same dimension and any other numerical invariants must be locally constant in a flat family. For example, if $X \xrightarrow{\pi} B$ is a flat family of projective varieties, meaning that each fiber is a projective variety, then all the members have the same degree and even the same Hilbert polynomial. In fact, it can be shown that a family of projective varieties over an irreducible base B is flat if and only if all members have the same Hilbert polynomial. Of course, there are many important examples of surjective morphisms that are not flat. In section 7.1, we will introduce blow-ups— surjective morphisms with the fiber over almost every point being just a single point, but the fiber over some special points being a large, complicated projective variety. Because the fibers vary so wildly, blowups are essentially never flat.

See [20, III 9] for more about flatness.

The preceding example of a hyperbola family has a nice interpretation in terms of scheme theory. The corresponding ring homomorphism between coordinate rings is the map

$$\mathbb{C}[t] \xrightarrow{\pi^\#} \frac{\mathbb{C}[x,y,z]}{(xy-z)},$$
$$t \longmapsto z.$$

The ring map $\pi^\#$ defines a map of schemes

$$\operatorname{Spec} \frac{\mathbb{C}[x,y,z]}{(xy-z)} \longrightarrow \operatorname{Spec}\mathbb{C}[t],$$
$$\mathfrak{p} \longmapsto (\pi^{\#})^{-1}(\mathfrak{p}),$$

whose restriction to maximal ideals recovers the original projection π.

The elements of $\operatorname{Spec}\mathbb{C}[t]$ are the prime ideals of the ring $\mathbb{C}[t]$. These are just the zero ideal (0) plus all the maximal ideals, which can be identified with the complex numbers. We have already encountered a surprising property of the nonmaximal prime ideal (0) of $\mathbb{C}[t]$ in the exercises: It is a *dense point* in the topological space $\operatorname{Spec}\mathbb{C}[t]$.

The fibers over points corresponding to complex numbers are the original hyperbolas, including the degenerate hyperbola over zero. It turns out that the fiber over the dense point is $\operatorname{Spec}\left(\frac{\mathbb{C}(z)[x,y]}{(xy-z)}\right)$, where $\mathbb{C}(z)$ is the field of complex rational functions—the fraction field of $\mathbb{C}[z]$. The indeterminate z here should now be treated as a constant in the ground field.

Thus the dense point also defines a hyperbola $\mathbb{V}(xy - z)$ in the plane $\mathbb{A}^2$ with coordinates x and y, but now the ground field is the field $\mathbb{C}(z)$ of rational functions in one variable. This "generic" hyperbola is smooth (as can be checked using the Jacobian criterion for smoothness described in the previous section), and this suggests that the set of smooth fibers is nonempty, and ought to be dense as well, since after all, this is a member of the family corresponding to a dense point of the parameter space.

6.4 Bertini's Theorem

The idea that a "general member" of a family of varieties ought to be smooth is also reflected in Bertini's theorem.

Before stating Bertini's theorem we recall that any hyperplane in $\mathbb{P}^n$ is the zero set of a linear functional $\sum_{i=0}^{n} a_i x_i$ on $\mathbb{C}^{n+1}$. The hyperplane determines the linear functional up to a nonzero multiplicative constant, so the hyperplane can be identified with a point $[a_0 : \ldots : a_n]$ in $\mathbb{P}^n$. Therefore, it is natural to think of the set of hyperplanes in $\mathbb{P}^n$ as the points in *dual projective space* $\check{\mathbb{P}}^n$.

Bertini's theorem: Let $V \subset \mathbb{P}^n$ be a smooth irreducible projective variety. Consider the set

$$W = \{H \in \check{\mathbb{P}}^n \mid H \cap V \text{ is a smooth variety }\}$$

of all hyperplanes in $\mathbb{P}^n$ intersecting V in a smooth variety. Then W is an open subvariety of $\check{\mathbb{P}}^n$. That is, the set of hyperplanes having singular intersection with V forms a Zariski-closed subset of $\check{\mathbb{P}}^n$.

As H varies through the dual space $\check{\mathbb{P}}^n$, the intersections $H \cap V$ form a family of varieties parametrized by $\check{\mathbb{P}}^n$, called the *hyperplane sections of V*.

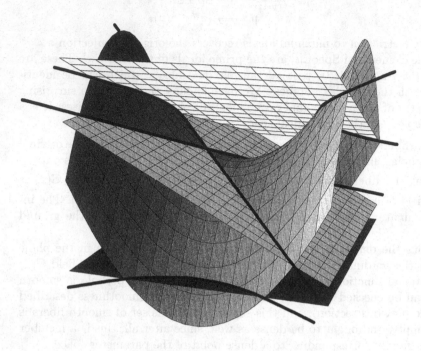

Figure 6.8. A general hyperplane section is smooth

Loosely, Bertini's theorem says that "the generic hyperplane section of V is smooth". For example, Figure 6.8 shows a few hyperplane sections of a fixed smooth variety. According to Bertini's theorem, a typical hyperplane section ought to be smooth, and it is easy to visualize that this is indeed true in this example. Indeed, a hyperplane section fails to be smooth if and only if the hyperplane is tangent to the variety at some point. For a proof of Bertini's theorem, see, for instance, [20, page 179].

Bertini's Theorem illustrates the general principle that a generic member of a family of algebraic varieties is smooth, provided that some member is smooth. Note that it is possible that no member of a family of algebraic varieties is smooth. Although the set of smooth members must be an open subset of the parametrizing variety, this open set can be empty. For example, consider the family of all tangent hyperplane sections of a fixed smooth variety. It is not very difficult to prove that no member of this family is smooth.

There are many variants of Bertini's theorem. For example, Figure 6.9 indicates a one-dimensional "linear" subfamily of the family of all hyperplane sections of a variety, namely, the family of hyperplane sections obtained by intersecting with planes parallel to a given plane. Because one such hyperplane section in this one-parameter family is smooth, a generic member of this family must be smooth.

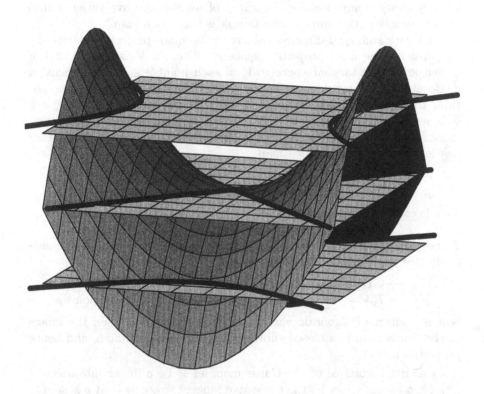

Figure 6.9. A one-dimensional subfamily of the family of all hyperplane sections

Note that Figure 6.9 suggests an alternate interpretation of the hyperbola family discussed in the preceding section. Instead of thinking of the members of the hyperbola family as the fibers of a map, we may think of them as hyperplane sections of the fixed variety in $\mathbb{A}^3$ defined by the vanishing of $z - xy$. We intersect only with the one-parameter family of hyperplanes given by $z = \lambda$ where λ varies. Refined versions of Bertini's theorem tell us that for generic λ, the corresponding member of the hyperbola family is a smooth variety. See [29] for a survey of the most recent progress in Bertini type theorems.

Exercise 6.4.1. Let V be the hypersurface in $\mathbb{A}^3$ defined by $z - xy = 0$. Show that a plane H in $\mathbb{A}^3$ is tangent to V if and only if the intersection $V \cap H$ is singular.

6.5   The Gauss Mapping

Using the fact that the Grassmannian is a projective algebraic variety, we can create many useful embeddings of quasi-projective varieties into projective space. One important example is the Gauss map.

Let V be a smooth d-dimensional irreducible quasi-projective variety sitting inside some fixed projective space $\mathbb{P}^n$. Because V is smooth and of dimension d, the tangent space to V at each point $p \in V$ has dimension d. Because we are considering V as subvariety of projective space, we consider the projective tangent space $T_pV \subset \mathbb{P}^n$. For each p in V, T_pV is a d-dimensional linear subspace of $\mathbb{P}^n$, that is, it is an element of the Grassmannian $\mathbf{Gr}(d+1, n+1)$ of d-dimensional linear subspaces of $\mathbb{P}^n$. In other words, we have a well-defined map

$$V \quad \longrightarrow \mathbf{Gr}(d+1, n+1),$$
$$p \quad \longmapsto T_pV = \text{ the projective tangent space to } V \text{ at the point } p.$$

This is called the *Gauss map* of V.

Theorem: If $V \subset \mathbb{P}^n$ is a smooth d-dimensional irreducible quasi-projective variety, then the Gauss map

$$V \quad \longrightarrow \mathbf{Gr}(d+1, n+1)$$
$$p \quad \longmapsto T_pV = \text{ the projective tangent space to } V \text{ at the point } p$$

is a morphism of algebraic varieties. If V is projective, then the image of the Gauss map is a closed subvariety of the Grassmannian, and hence projective as well.

As a trivial example of the Gauss map, let V be a linear subvariety of $\mathbb{P}^n$. Then for each p in V, the projective tangent space to V at p is simply V itself. (The tangent space is, after all, the linear subvariety that best approximates V.) Thus for every p in V, the Gauss map associates the point in the Grassmannian corresponding to V itself. In other words, the image of the Gauss map for a linear variety is a single point. The case of a linear subvariety, however, is quite exceptional, as the next theorem shows.

Theorem: The image of the Gauss map is a quasi-projective variety of the same dimension as the original variety, unless the original variety is a linear subvariety of projective space (in which case the dimension of the image of the Gauss map is zero).

Harris [17, page 188] gives further information about the Gauss map. For a proof of the above theorem, see [15].

One special case of the Gauss map was extensively studied in the nineteenth century: The case where the variety is a plane curve.

If $C \subseteq \mathbb{P}^2$ is a smooth plane curve, then the tangent space to each point of C is a line in $\mathbb{P}^2$. On the other hand, the lines in $\mathbb{P}^2$ can be identified with another copy of $\mathbb{P}^2$, called the *dual projective plane* and denoted $\check{\mathbb{P}}^2$. The correspondence is simple: Each line in $\mathbb{P}^2$ is the zero set of a linear polynomial $a_0 x_0 + a_1 x_1 + a_2 x_2$, where a_0, a_1, a_2 are complex numbers, not all zero. Two such triples (a_0, a_1, a_2) and (b_0, b_1, b_2) determine the same line if and only if they are multiples of each other, which is to say that the line $\mathbb{V}(a_0 x_0 + a_1 x_1 + a_2 x_2)$ in $\mathbb{P}^2$ corresponds uniquely to the point $[a_0 : a_1 : a_2] \in \check{\mathbb{P}}^2$ determined by the coefficients of its defining equation. Therefore, the set of lines in $\mathbb{P}^2$, $\mathbf{Gr}(2,3)$, is identified with the dual projective plane $\check{\mathbb{P}}^2$.

Now, the Gauss map can be expressed as a map

$$C \longrightarrow \check{\mathbb{P}}^2,$$
$$x \longmapsto \text{(coefficients of the defining equation of the tangent line to } x\text{)}.$$

By the theorem above, the image of this map is again a curve in $\check{\mathbb{P}}^2$, called the *dual curve*. Studying the dual curve can sometimes reveal interesting properties of the original curve C.

Exercise 6.5.1. Convince yourself that the image of the Gauss map in the case of a plane curve is indeed a curve in $\check{\mathbb{P}}^2$.

Exercise 6.5.2. Study the Gauss map for a plane curve given by $y = x^3$. What geometric feature of this curve causes singularities in the dual curve?

Exercise 6.5.3. Let $V \subseteq \mathbb{P}^2$ be the curve defined by the equation $x^d + y^d + z^d = 0$, for $d \geq 2$. Describe the Gauss map $V \to \check{\mathbb{P}}^2$ explicitly. (Hint: Use elementary calculus to find the tangent line to V).[3] Find the defining equation of the dual curve when $d = 2$ and when $d = 3$. Is the dual curve smooth in general?

Exercise 6.5.4. Show that the Gauss map for a projective variety V depends on the choice of the embedding of V in projective space by comparing the Gauss map for a line and for a conic in the projective plane.

Exercise 6.5.5. Prove that the hyperplanes in $\mathbb{P}^n$ are in one-to-one correspondence with the points in another copy of $\mathbb{P}^n$, called the *dual projective space,* and denoted by $(\mathbb{P}^n)\check{}$. More generally, prove that the hypersurfaces in $\mathbb{P}^n$ of degree d are in one-to-one correspondence with the points in a projective space of dimension $\binom{n+d}{d} - 1$. (Hint: Revisit Section 5.2 on conics.)

---

[3]Solution: $[x : y : z] \mapsto [x^{d-1} : y^{d-1} : z^{d-1}]$.

Exercise 6.5.6. Using Exercise 6.5.3, prove that there are exactly four lines tangent to two distinct non-degenerate plane conics (counting multiplicities). What is the geometric meaning of "multiplicity" in this context?

7

Birational Geometry

7.1 Resolution of Singularities

In 1964, Heisuke Hironaka proved a fundamental theorem: Every quasi-projective variety can be *desingularized*, or equivalently, every variety is "birationally equivalent" to a smooth projective variety. Before we can state this theorem, we need to introduce some new ideas.

Definition: A morphism of varieties $X \xrightarrow{\pi} V$ is called a *projective morphism*[1] if X is a closed subvariety of a product variety

$$X \subset V \times \mathbb{P}^n$$

and $X \xrightarrow{\pi} V$ is the restriction of the projection onto the first coordinate.

Projective morphisms have the property that the preimage of any point is a projective variety. A projective morphism is a *proper mapping* in the Euclidean topology, that is to say, the preimage of any compact set is compact in the Euclidean topology.

Definition: A morphism $X \xrightarrow{\pi} V$ of quasi-projective varieties is called a *birational morphism* if its restriction to some dense open set $U \subset X$ is an isomorphism onto some dense open subset $U' \subset V$.

A birational morphism need not be one-to-one, nor surjective.

---

[1]Do not confuse a projective morphism with a morphism of projective varieties.

Hironaka's Desingularization Theorem: Let V be a quasi-projective variety. Then there exists a smooth quasi-projective variety X and a projective birational morphism $X \xrightarrow{\pi} V$. Furthermore, π may be assumed to be an isomorphism on the smooth locus of V, and if V is a projective variety, then so is X.

The theorem says that every quasi-projective variety V, no matter how badly singular, admits a *desingularization* X, which is a smooth variety projecting to V that looks just like V everywhere except at the singular points of V. When we say that π is an isomorphism on the smooth locus of V we mean that π restricts to an isomorphism of varieties on the dense open sets

$$X \smallsetminus \pi^{-1}(\text{Sing } V) \xrightarrow{\pi} V \smallsetminus \text{Sing } V.$$

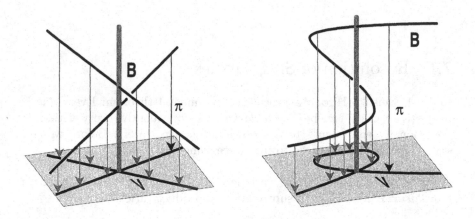

Figure 7.1. Two desingularizations

On the proof: The case where V is one-dimensional, the curve case, is relatively easy and requires no restrictions on the ground field. It follows from the algebraic process called "normalization" (see [37, Chapter II, Section 5.4]). The case of surfaces was understood by the Italian school of early-twentieth-century algebraic geometers and can be understood by students of algebraic geometry without too much difficulty (see [2]). However, by modern standards it may be argued that they did not really have a proof for the general case of Hironaka's theorem for abstract algebraic surfaces. In the forties, Zariski gave a "purely algebraic" proof of resolution of singularities for any algebraic surface or three-fold defined over the complex numbers, or over any field of characteristic 0. Abhyankar later established resolution for surfaces and three-folds of nonzero characteristic.

Hironaka's proof of the higher-dimensional case is quite difficult, spanning two issues of the *Annals of Mathematics* [21], and is valid only for varieties defined over a field of characteristic zero. This beautiful and fundamental work was recognized with a Fields medal in 1970. Although one might argue that in theory Hironaka's theorem is algorithmic, in practice it is virtually impossible to use it to construct a resolution of singularities for a given variety. Recently, Villamayor [39] and, independently, Bierstone and Milman [3] have clarified the process of resolution of singularities in characteristic zero, explicitly describing the algorithmic nature of the resolution process. Lipman's review of Bierstone and Milman's paper [31] surveys this history and describes recent developments in the problem of resolution of singularities, including numerous references to the literature.

Recently, exciting developments have grown from the deep ideas of Johan de Jong, who proved that projective maps exist in very general situations that are as good as desingularizations for many purposes. Using de Jong's ideas, simple proofs of Hironaka's theorem have been discovered by de Jong and Abramovich [1] and by Bogomolov and Pantev [4].

$$\square$$

Hironaka's theorem is founded on a very concrete construction, called *blowing up*. Let us look at some examples of blowups. To start, we will forget about singularities and just describe the blowup of a point in affine space.

In blowing up $\mathbb{A}^n$ at a point p, the idea is to leave $\mathbb{A}^n$ unaltered except at the point p, which is replaced by the set of all lines through p, a copy of $\mathbb{P}^{n-1}$. To make this precise, let us choose a suitable coordinate system for $\mathbb{A}^n$ so that the point p may be assumed to be the "origin." Let B be the set of all pairs (x, ℓ), where $x \in \mathbb{A}^n$ and $\ell \in \mathbb{P}^{n-1}$ is a line through the origin of $\mathbb{A}^n$ containing x. That is,

$$B = \left\{ (x, \ell) \in \mathbb{A}^n \times \mathbb{P}^{n-1} \mid x \in \ell \right\} \subset \mathbb{A}^n \times \mathbb{P}^{n-1}.$$

The blowup of $\mathbb{A}^n$ at p is by definition the natural projection to the affine factor

$$
\begin{aligned}
B & \xrightarrow{\ \pi\ } \mathbb{A}^n, \\
(x, \ell) & \longmapsto x.
\end{aligned}
$$

Let us check that the blowup $B \longrightarrow \mathbb{A}^n$ has the desired properties by considering the fibers of this map. The fiber of π over any point x other than the origin is simply the single point (x, ℓ), where ℓ is the unique line through x and the origin. However, the fiber over the origin p is an entire copy of $\mathbb{P}^{n-1}$, namely $\left\{ (p, \mathbb{P}^{n-1}) \right\} \subset \mathbb{A}^n \times \mathbb{P}^{n-1}$, since the origin lies on *every* line through the origin. The blowing up morphism $B \longrightarrow \mathbb{A}^n$ collapses this $\mathbb{P}^{n-1}$ to a point and is bijective everywhere else. See Figure 7.2.

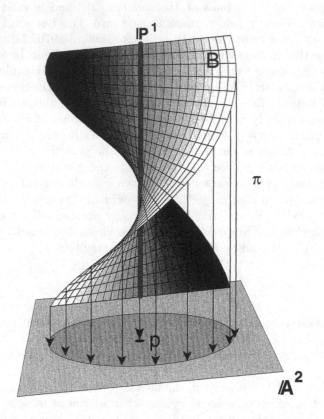

Figure 7.2. Blowup of the plane at a point

We claim that the set B is a quasi-projective variety. Indeed, if $(x_1, \ldots, x_n)$ are the coordinates of $\mathbb{A}^n$ and $[y_1 : \cdots : y_n]$ are the coordinates of $\mathbb{P}^{n-1}$, then $(x_1, \ldots, x_n; y_1 : \cdots : y_n)$ are coordinates for $\mathbb{A}^n \times \mathbb{P}^{n-1}$. Now, a point $x = (x_1, \ldots, x_n) \in \mathbb{A}^n$ lies on the line ℓ represented by $[y_1 : \ldots : y_n]$ in $\mathbb{P}^{n-1}$ if and only if the vector $(x_1, \ldots, x_n)$ is a multiple (possibly 0) of the vector $(y_1, \ldots, y_n)$. That is, x lies on ℓ if and only if the matrix

$$\begin{bmatrix} x_1 & \cdots & x_n \\ y_1 & \cdots & y_n \end{bmatrix}$$

has rank less than or equal to 1. This holds precisely when all the 2×2 minors of this matrix vanish. That is, the point $x = (x_1, \ldots, x_n)$ lies on

$\ell = [y_1 : \cdots : y_n]$ if and only if the coordinates $(x_1, \ldots, x_n; y_1 : \cdots : y_n)$ satisfy the polynomial equations $x_i y_j - x_j y_i = 0$ for all i and j. Thus

$$B = \mathbb{V}(x_i y_j - x_j y_i \mid 0 \leq i < j \leq n) \subseteq \mathbb{A}^n \times \mathbb{P}^{n-1}.$$

The reader should have no trouble checking that such a zero set is indeed a quasi-projective variety (however, B is neither affine nor projective!). Indeed, B is a closed subset in the quasi-projective variety $\mathbb{A}^n \times \mathbb{P}^{n-1}$.

The blowing up morphism $B \xrightarrow{\pi} \mathbb{A}^n$ is obviously a projective, birational morphism. Indeed, since B is a closed subvariety of $\mathbb{A}^n \times \mathbb{P}^{n-1}$ and π is the restriction of the natural projection to $\mathbb{A}^n$, the map π is projective by definition. Furthermore, the map

$$\mathbb{A}^n \smallsetminus \{p\} \quad \to B \subseteq \mathbb{A}^n \times \mathbb{P}^{n-1},$$
$$(x_1, \ldots, x_n) \quad \longmapsto (x_1, \ldots, x_n; x_1 : \cdots : x_n),$$

is easily seen to be a morphism of quasi-projective varieties inverse to π on the dense open set $\mathbb{A}^n \smallsetminus \{p\}$.

The variety B, together with its natural projection $B \xrightarrow{\pi} \mathbb{A}^n$, is sometimes called the *one-point blowup of* $\mathbb{A}^n$. We also denote the space B by $B_p(\mathbb{A}^n)$. We think of this variety as obtained by removing the origin from $\mathbb{A}^n$ and replacing it by the set of all lines through p in $\mathbb{A}^n$.

We will later treat the projection onto the other factor

$$B \xrightarrow{\pi_2} \mathbb{P}^{n-1},$$
$$(x, \ell) \longmapsto \ell,$$

which makes B into a *line bundle* over projective space $\mathbb{P}^{n-1}$, called the *tautological line bundle* over $\mathbb{P}^{n-1}$.

The blowup of affine space at a point can be generalized in several directions. We can blow up a larger subvariety of $\mathbb{A}^n$ instead of a point, or we can blow up points in a more general quasi-projective variety. Ultimately, we would like to blow up arbitrary closed subvarieties of an arbitrary quasi-projective variety (and even closed subschemes of an arbitrary scheme!). We start with the blowup of a point in an arbitrary affine variety.

Definition: Let $V \subset \mathbb{A}^n$ be an affine algebraic variety and p a point of V. The *blowup* of V at p is the Zariski closure of the preimage

$$\pi^{-1}(V \smallsetminus \{p\})$$

in the variety B obtained by blowing up p in $\mathbb{A}^n$, together with the natural projection π to V. We denote the blowup of V at p by $B_p(V)$.

Because $B_p(\mathbb{A}^n) \xrightarrow{\pi} \mathbb{A}^n$ is an isomorphism when restricted to the open set $B_p(\mathbb{A}^n) \smallsetminus \pi^{-1}(p)$, the restriction of π to $B_p(V) \smallsetminus \pi^{-1}(p)$ is an isomorphism onto $V \smallsetminus \{p\}$. An example of such a blowup is illustrated in Figure 7.3 on the following page.

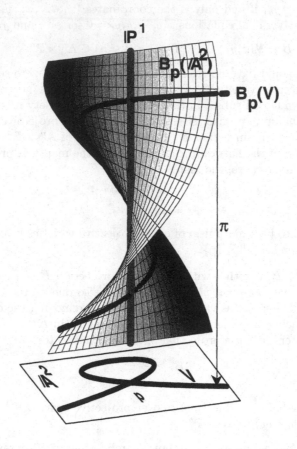

Figure 7.3. Blowup of a nodal curve at the nodal point

Example: Let us blow up the cone $\mathbb{V}(x^2 + y^2 - z^2) \subset \mathbb{A}^3$ at the origin. As illustrated in Figure 7.4, the cone looks like a union of lines, tied together at the origin.

The singularity of the cone is a result of this tying together of lines. To desingularize the cone, we must separate these lines. Intuitively, the desingularization of the cone is the *disjoint* union of the lines, in other words, a cylinder. The desingularizing process is depicted in Figure 7.5.

The "origins" of the lines have been separated, and they correspond to a circle on the cylinder.

We now work this out explicitly by blowing up. The blowup map is

$$B = \{(x, \ell) \in \mathbb{A}^3 \times \mathbb{P}^2 \mid x \in \ell\} \;\xrightarrow{\;\pi\;}\; \mathbb{A}^3,$$
$$(x, \ell) \longmapsto x.$$

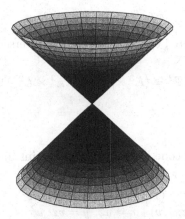

Figure 7.4. Lines tied together at a point form a cone

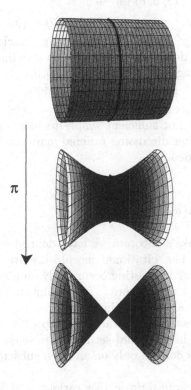

Figure 7.5. Desingularizing the cone

Because projective space $\mathbb{P}^2$ can be covered by its three standard charts

$$\mathbb{P}^2 = \mathbb{A}^2{}_x \cup \mathbb{A}^2{}_y \cup \mathbb{A}^2{}_z,$$

the blowup is also covered by three charts: The intersection with each chart

$$B \subset \mathbb{A}^3 \times \mathbb{P}^2 \cong (\mathbb{A}^3 \times \mathbb{A}^2{}_x) \cup (\mathbb{A}^3 \times \mathbb{A}^2{}_y) \cup (\mathbb{A}^3 \times \mathbb{A}^2{}_z)$$

$$\pi \downarrow$$

$$\mathbb{A}^3$$

of the preimage of the cone is easily found. For instance, the last one is

$$
\begin{aligned}
V &= \pi^{-1}(\mathbb{V}(x^2 + y^2 - z^2)) \cap (\mathbb{A}^3 \times \mathbb{A}^2{}_z) \\
&= \left\{ ((x,y,z),(x:y:z)) \mid x^2 + y^2 = z^2 \right\} \\
&\cong \left\{ (x,y,z,u,v) \mid x = uz,\, y = vz,\, u^2 + v^2 = 1 \right\} \subset \mathbb{A}^5.
\end{aligned}
$$

Projecting $\mathbb{A}^5$ into 3-dimensional space $\mathbb{A}^3$, with coordinates z, u, and v the image set

$$\left\{ (z,u,v) \mid u^2 + v^2 = 1 \right\}$$

is a variety isomorphic to V. This is a cylinder in $\mathbb{A}^3$. We have blown up the vertex of a cone $\mathbb{V}(x^2 + y^2 - z^2) \subset \mathbb{A}^3$, producing a variety $B \subset \mathbb{A}^3 \times \mathbb{P}^2$, which contains an open dense set isomorphic to a cylinder. Under these identifications, the preimage of the vertex point $(0,0,0) \in \mathbb{V}(x^2 + y^2 - z^2) \subset \mathbb{A}^3$ is the circle $\{z = 0, u^2 + v^2 = 1\}$ on the cylinder.

To desingularize more complicated varieties, we need to blow up more general subvarieties: It is not sufficient simply to blow up points. We will return to this problem after discussing rational maps, on which the idea of a general blowup is founded.

7.2 Rational Maps

In the context of Hironaka's theorem we have defined what we mean by a birational morphism. The birational morphisms are not well defined everywhere, but everything interesting occurs only on a nonempty Zariski-open set, that is, "almost everywhere." The definition of a rational map formalizes this "almost everywhere" thinking.

The first thing to stress about a rational map of a variety X is that it is not actually a map in the usual set-theoretic sense. Rather, it is an equivalence class of maps defined only on an open subset of X.

Definition: Let X be a quasi-projective variety, and let U and U' be dense open subsets of X. Suppose we are given two morphisms $U \xrightarrow{\varphi} Y$ and $U' \xrightarrow{\varphi'} Y$ of quasi-projective varieties. We say that (U, φ) and (U', φ')

are *equivalent* if the mappings φ and φ' coincide on the intersection $U \cap U'$. It is easy to check that this forms an equivalence relation.

Definition: A *rational map* $X \dashrightarrow Y$ is an equivalence class of morphisms defined on dense open subsets of X as above.

We think of a rational map as a morphism defined only on a dense open set, and we do not concern ourselves with the particular open set on which it is defined. Nevertheless, a rational map has a unique extremal *domain of definition*. Since the domain of definition of every representative is dense, inspecting the domain of definition of the rational map is usually superfluous. On the domain of definition (which, as it turns out, is open) the rational map is a morphism of varieties. Thus a rational map is a "morphism defined almost everywhere."

A rational map, despite its name, is not an actual mapping, which is the reason we use a broken arrow to denote it. This can lead to trouble, for example, when composing two rational maps. One must take care that the image of (a representative of) φ_1 is dense in Y in order to define a composition

$$X \xdashrightarrow{\varphi_1} Y \xdashrightarrow{\varphi_2} Z.$$

Example: Projection from a point in projective space is an example of a rational map. Let $H \subset \mathbb{P}^n$ be a fixed hyperplane in $\mathbb{P}^n$ and let $p \in \mathbb{P}^n$ be any point not on H. The *projection* from p onto the hyperplane H is the rational map

$$\mathbb{P}^n \xdashrightarrow{\varphi} H = \mathbb{P}^{n-1},$$
$$x \longmapsto \varphi(x) = \text{the unique intersection point of } H \text{ and the line } \overline{xp}.$$

The rational map is a well-defined morphism everywhere outside p.

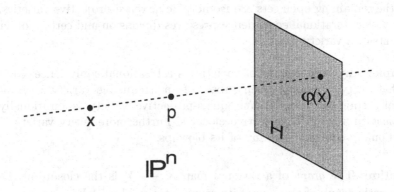

Figure 7.6. Projection from p onto H

In homogeneous coordinates the projection is most easily expressed by choosing the origin and axes in such a way that $p = [0 : \ldots : 0 : 1]$ and $H = \mathbb{V}(x_n) \subset \mathbb{P}^n$ is identified with a copy of $\mathbb{P}^{n-1}$. Then the projection from p takes the form

$$\mathbb{P}^n \dashrightarrow^{\varphi} \mathbb{P}^{n-1},$$
$$[x_0 : \ldots : x_n] \longmapsto [x_0 : \ldots : x_{n-1}] = \varphi(x).$$

7.3 Birational Equivalence

Definition: Let X and Y be irreducible algebraic varieties. We call them *birationally equivalent* if there are mutually inverse rational maps

$$X \dashrightarrow^{F} Y$$

and

$$Y \dashrightarrow^{G} X.$$

This means that the compositions $F \circ G$ and $G \circ F$ are defined and that each is the identity rational map, that is, both $F \circ G$ and $G \circ F$ agree with the identity map on dense open sets where the compositions make sense as morphisms of algebraic varieties.

Basically, X and Y are birationally equivalent if they are isomorphic on a (dense) open set. That is, after removing some closed subvariety of X and of Y, the remaining open sets are isomorphic as quasi-projective varieties. In particular, birational equivalence preserves dimension and certain other invariants of a variety.

Example: Any isomorphism of varieties is a birational equivalence. Note also that every nonempty open subset of an irreducible variety V is birationally equivalent to V. Any quasi-projective variety is birationally equivalent to any of its projective closures. Furthermore, every variety V is birationally equivalent to any of its blowups.

Definition: The *graph* of a rational map $X \dashrightarrow^{F} Y$ is the closure of the set-theoretic graph of any one of its representatives $U \xrightarrow{\varphi} Y$:

$$\Gamma_F = \overline{\{(x, \varphi(x)) \mid x \in U\}} \subset X \times Y.$$

The closure is taken in the Zariski topology of the product variety $X \times Y$, but we get the same result by taking the closure in the Euclidean topology on the product $X \times Y$. The reader should check that the closure is well-defined: graphs of different representatives all have the same closure.

Exercise 7.3.1. Let Γ be the graph of a rational map $X \dashrightarrow Y$. Prove that the projection $\Gamma \longrightarrow X$ is a birational equivalence.

Exercise 7.3.2. The *function field* of an irreducible affine variety is defined as the fraction field of its coordinate ring. More generally, the function field of a nonaffine irreducible variety is defined as the function field of any nonempty affine open subset. Show that this is independent of the choice of the affine open set.

Exercise 7.3.3. Prove that two irreducible varieties X and Y are birationally equivalent if and only if their function fields $\mathbb{C}(X)$ and $\mathbb{C}(Y)$ are isomorphic as $\mathbb{C}$-algebras.

Exercise 7.3.4. Prove that $\mathbb{P}^2$ is birationally equivalent, but not isomorphic, to $\mathbb{P}^1 \times \mathbb{P}^1$.

Exercise 7.3.5. Find the equation defining the graph of the rational map $\mathbb{A}^2 \dashrightarrow \mathbb{A}^1$ sending (x, y) to y/x as a subvariety of $\mathbb{A}^3$.

7.4   Blowing Up Along an Ideal

Now we can define the blowup of more general subvarieties. To facilitate our understanding, we introduce an alternative interpretation of blowing up a point in $\mathbb{A}^n$.

Consider the map

$$\mathbb{A}^n \smallsetminus \{0\} \xrightarrow{\ell} \mathbb{P}^{n-1},$$
$$x = (x_1, \ldots, x_n) \longmapsto \ell(x) = [x_1 : \ldots : x_n],$$

attaching to each point $x \in \mathbb{A}^n \smallsetminus \{0\}$ the line $\ell(x)$ through 0 and x.

As a set, the graph of this map is

$$\{(x, \ell(x)) \in \mathbb{A}^n \smallsetminus \{0\} \times \mathbb{P}^{n-1}\}.$$

It is not hard to see that the blowup $B_p(\mathbb{A}^n)$ is the closure of this graph in the product variety $\mathbb{A}^n \times \mathbb{P}^{n-1}$. Thus, the blowup of the origin in $\mathbb{A}^n$ can be identified with the graph of the rational map

$$\mathbb{A}^n \dashrightarrow \mathbb{P}^{n-1}$$
$$(x_1, \ldots, x_n) \longmapsto [x_1 : \cdots : x_n].$$

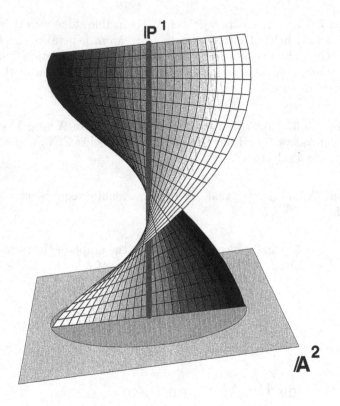

Figure 7.7. Graph in $\mathbb{A}^2 \times \mathbb{P}^1$ of the rational map $\mathbb{A}^2 \dashrightarrow \mathbb{P}^1$

We use this idea to blow up an arbitrary subvariety Y of a variety X.

Definition: Let $F_1, \ldots, F_r$ be functions in the coordinate ring $\mathbb{C}[X]$ of an irreducible affine algebraic variety X, and let I be the ideal they generate. Assume that I is a proper non-zero ideal of $\mathbb{C}[X]$. The *blowup* of the variety X along the ideal I is the graph B of the rational map

$$X \xrightarrow{F} \mathbb{P}^{r-1},$$
$$x \longmapsto [F_1(x) : \ldots : F_r(x)],$$

together with the natural projection map $B \subset X \times \mathbb{P}^{r-1} \xrightarrow{\pi} X$. The blowup of X along I is denoted $B_I(X)$.

The projection

$$B_I(X) \xrightarrow{\pi} X,$$
$$(x, F(x)) \longmapsto x,$$

defines an isomorphism of quasi-projective varieties between the open sets

$$B_I(X) \smallsetminus \pi^{-1}(Y) \to X \smallsetminus Y,$$

where Y is the closed set in X defined by the vanishing of $F_1, \ldots, F_r$. Indeed, the inverse morphism can be defined as

$$X \smallsetminus Y \longrightarrow B_I(X) \subseteq X \times \mathbb{P}^{r-1},$$
$$x \longmapsto (x, [F_1(x) : \cdots : F_r(x)]),$$

which is well-defined on $X \smallsetminus Y$ because the functions $F_1, \ldots, F_r$ do not simultaneously vanish anywhere on $X \smallsetminus Y$. In other words, the rational map

$$X \dashrightarrow B_I(X),$$
$$x \longmapsto (x, [F(x)]),$$

is an inverse of the blowing up map $B_I(X) \xrightarrow{\pi} X$, showing that X and $B_I(X)$ are birationally equivalent varieties.

Although it is not obvious, the isomorphism class of the blowup $B_I(X)$ depends only on the ideal I and not on the particular choice of generators. Furthermore, if I happens to be the maximal ideal corresponding to a point x in X, then the blowup $B_I(X)$ agrees with the the blowup of X at the point x defined in Section 7.1. For more on blowing up from this point of view, see [10, Section IV.2].

We have not made any assumptions as to whether the ideal is radical or not. Different ideals may produce different blowups, even if both have the same radical, which is to say, even if they define the same closed subset of X. This means that we are really blowing up along an ideal, not just the subvariety defined by that ideal. Furthermore, it is not really necessary to assume that X is irreducible; if X has several components, one should assume that the functions F_i do not all vanish along any of them in order for the above discussion to be valid.

Definition: Let Y be an irreducible subvariety of an affine algebraic variety X. The *blowup of X along the subvariety Y* is the blowup along the radical ideal $\mathbb{I}(Y)$. We also denote this by $B_Y(X)$.

So far, we have considered only affine varieties, but this restriction is not necessary. The discussion also makes sense for a quasi-projective variety $X \subset \mathbb{P}^n$. For example, we can take a homogeneous ideal, generated by homogeneous polynomials $F_1, \ldots, F_r$ of the same degree, and consider the rational map

$$X \xrightarrow{F} \mathbb{P}^{r-1},$$
$$x \longmapsto [F_1(x) : \ldots : F_r(x)],$$

The graph of the rational map F (together with its projection to X) is the *blowup* of X along the ideal $(F_1, \ldots, F_r)$ in X. Blowing up along an arbitrary ideal that is not necessarily radical is also called *blowing up along the subscheme* defined by the ideal I.

As the reader might suspect, it is in fact possible to blow up along any subvariety (or subscheme) in any variety. To do this correctly, one should introduce the machinery of an ideal sheaf $\mathcal{I}$ in the structure sheaf $\mathcal{O}_X$ of the variety, and define the blowup of X along the ideal sheaf $\mathcal{I}$ by patching together the blowups on affine charts of X (see [20, Chapter II, Section 7]).

As we have seen, every blowup morphism $\tilde{X} \to X$ is a projective birational map. In fact, although it is not obvious, the converse is also true: Every projective birational map $\tilde{X} \to X$ of quasi-projective varieties is a blowup of some sheaf of ideals in X (see [20, Theorem II.7.17]).

We return to Hironaka's desingularization theorem. Let V be an affine variety; say V is a Zariski-closed set in $\mathbb{A}^n$. Hironaka's theorem says that there exists a set of polynomials $F_0, \ldots, F_r$ in n variables such that the graph of the rational map

$$V \dashrightarrow \mathbb{P}^r,$$
$$x \longmapsto [F_0(x) : \cdots : F_r(x)]$$

is a desingularization of V. Denoting this graph by X, we know that X is a closed subvariety of $V \times \mathbb{P}^r$, and the natural projection $X \xrightarrow{\pi} V$ onto the first factor is a birational equivalence, called the blowup of V along the ideal $(F_0, \ldots, F_r)$.

The remarkable fact proved by Hironaka is that the F_i can be chosen so that the variety X is *smooth*. Furthermore, Hironaka's theorem guarantees that the F_i can be chosen so that the closed subset $\mathbb{V}(F_0, \ldots, F_r) \subseteq V$ where the map fails to be an isomorphism is precisely the singular locus of V. Because the projection $X \to V$ is an isomorphism on the dense open set complementary to the singular locus, we see that X looks "just like V" except on the singular set of V.

It is important to realize that the ideal $(F_0, \ldots, F_r)$ may not be the same as the Jacobian ideal defined in Section 6.2, although both ideals define the singular locus of V. The fact that they define the same closed set of V means only that they must have the same radical. Indeed, although we know that a desingularizing ideal exists, it is difficult to identify one explicitly.

The statement of Hironaka's theorem can be refined as follows. Rather than blowing up the nonradical ideal $(F_0, \ldots, F_r)$, a variety can be desingularized by successively blowing up radical ideals that define smooth subvarieties contained in the singular locus. Sometimes this is more useful in practice, because blowing up a smooth subvariety has a geometric interpretation similar to the interpretation of blowing up a point. Instead of replacing the point by the set of all lines through that point, blowing up Y replaces Y by all *normal directions* to Y. To make this precise requires more machinery than we develop here; see [37, Book II Chapter VI, Section 2].

Hironaka's theorem does not claim that the desingularization is unique or in any way canonical. Every variety (of dimension greater than one) has many nonisomorphic desingularizations. In Section 7.6 we discuss some of the research that has been done on the question of whether there is some sort of canonical or "minimal" smooth projective variety that is isomorphic to a given variety on a dense open set.

Exercise 7.4.1. Show that the variety obtained by blowing up a maximal ideal in an affine variety is the same as the variety obtained by blowing up any power of that ideal. (Hint: Recall the Veronese mapping.)

Exercise 7.4.2. Let $X \subset \mathbb{A}^{n+1}$ be the affine cone over a smooth projective variety. Show that the cone X can be desingularized by blowing up its vertex. What is the fiber of the desingularizing map over the vertex?

7.5 Hypersurfaces

Our goal in this section is to explain why every irreducible projective variety is birationally equivalent to a hypersurface. In other words, given any irreducible projective variety V of dimension d, there exists a hypersurface

$$X = \mathbb{V}(F) \subseteq \mathbb{P}^{d+1}$$

such that V and X contain isomorphic dense open sets. The easiest way to prove this uses a purely algebraic argument in field theory; see [20, page 27]. However, we would like to sketch a more intuitive geometric argument. This will complete the proof of the nonemptiness of the smooth locus as promised in Section 6.2.

Sketch of the proof: Let $V \subseteq \mathbb{P}^n$ be an irreducible projective variety. We first substantiate the claim that if V has dimension $n - 1$, then V must be defined by a single equation; that is, every codimension-one irreducible subvariety of $\mathbb{P}^n$ is a hypersurface.

To see this, first note that the ideal $\mathbb{I}(V)$ of functions vanishing on V must contain some *irreducible* homogeneous polynomial F. Indeed, take any F in $\mathbb{I}(V)$. If it factors as GH, then because $\mathbb{I}(V)$ is prime, G or H must be in $\mathbb{I}(V)$. By induction on degree, eventually $\mathbb{I}(V)$ contains some homogeneous irreducible polynomial. Using the fact that every polynomial factors uniquely (up to units) into irreducible factors, one checks easily that the ideal generated by an irreducible polynomial is prime, so that we have an inclusion of prime ideals

$$(F) \subset \mathbb{I}(V) \subsetneq \mathbb{C}[x_0, \ldots, x_n].$$

If $(F) \neq \mathbb{I}(V)$, then V is a proper subvariety of the irreducible codimension one subvariety $\mathbb{V}(F)$, and hence must have codimension at least two. This contradiction ensures that $\mathbb{I}(V) = (F)$, and so $V = \mathbb{V}(\mathbb{I}(V)) = \mathbb{V}(F)$. Thus every irreducible codimension-one subvariety of $\mathbb{P}^n$ is a hypersurface.

The proof of the theorem can now be completed by induction on the codimension of V in $\mathbb{P}^n$. We have just established the codimension-one case.

Suppose codim $V > 1$. Fix a point $p \in \mathbb{P}^n \setminus V$ and a hyperplane $H \subseteq \mathbb{P}^n$ not containing p. Let π be the projection from p onto H:

$$\mathbb{P}^n \overset{\pi}{\dashrightarrow} H \cong \mathbb{P}^{n-1},$$
$$x \longmapsto \pi(x),$$

as defined in Section 7.2. Let $V' \subseteq \mathbb{P}^{n-1}$ be the image of V under this map. It is easy to see that p and H can be chosen so that the projection

$$V \overset{\pi}{\longrightarrow} V'$$

is one-to-one on a dense open set. Then an inverse map can be constructed on a dense open set, and one easily verifies that this defines a birational equivalence. Because the codimension of V' in $\mathbb{P}^{n-1}$ is one less than the codimension of V in $\mathbb{P}^n$, the proof is complete by induction on the codimension.   □

While it is intuitively obvious that the projection $V \overset{\pi}{\longrightarrow} V'$ is a birational equivalence for generic choices of p and H, the reader may have trouble finding a precise geometric proof. This should help one appreciate the simple algebraic proof given in Hartshorne's book, which, however, requires a certain amount of field theory.

Exercise 7.5.1. An algebraic variety is said to be *rational* if it is birationally equivalent to projective space (of some dimension). Show that the nodal plane curve defined by the equation $y^2 - x^2 - x^3 = 0$ is rational. (Hint: Project from the node.)

7.6 The Classification Problems

We start with a list of guiding problems that motivate much of the research in algebraic geometry. For the most part, these problems are hopelessly difficult to answer in general, but progress can be measured against them.

- Classify all varieties up to isomorphism.

- Classify all smooth projective varieties up to birational equivalence. Notice that this would give a birational classification of all quasi-projective varieties, since every quasi-projective variety is birationally equivalent to a projective variety and, by Hironaka's theorem, every projective variety is birationally equivalent to a smooth projective variety. This problem is equivalent to the purely algebraic problem of classifying finitely generated field extensions of $\mathbb{C}$ up to isomorphism.

- Classify the varieties in each birational equivalence class up to isomorphism.

- Choose a canonical representative for each birational equivalence class.

For curves (one-dimensional varieties) all of these questions have satisfactory answers, which have been developed during centuries of beautiful mathematics.

We summarize the theory for curves without indicating the proofs. First, it is relatively easy to prove that each birational equivalence class has a unique smooth projective model (see [20, page 45]). Furthermore, every rational map between smooth projective curves extends uniformly to a well-defined morphism; hence birational maps and isomorphisms are the same for smooth projective curves. Because smooth complex projective curves are compact Riemann surfaces, classifying complex curves has led to an algebraic analogue of *Teichmüller theory,* which studies the moduli of Riemann surfaces up to conformal isomorphism. From the viewpoint of algebraic geometry the main results about smooth projective curves are as follows

- There exists only one smooth projective genus-zero curve up to isomorphism, namely $\mathbb{P}^1$.

- There exists a one-parameter family of isomorphism classes of smooth projective curves of genus one, the so-called *elliptic curves,* indexed by the *j-invariant,* a parameter varying over $\mathbb{A}^1$ (see [20, Chapter IV, Section 4]).

- The smooth projective curves of genus greater than one are parametrized by the *moduli spaces* $\mathfrak{M}_g$. These moduli spaces were first constructed by David Mumford as abstract $(3g - 3)$-dimensional varieties[2] [33] but soon afterward were shown to be, in fact, irreducible quasi-projective varieties by Deligne and Mumford [8]. The structure of these moduli spaces and their generalizations is an active field of research, especially since interesting connections with theoretical physics were discovered in the past ten years by Witten, Kontsevich, and others; see [18].

---

[2]Abstract varieties are defined in the Appendix.

As the above summary indicates, quite a lot is known about the classification of curves. Nonetheless, questions still abound. For example, although we will prove in the next section that every smooth projective curve can be embedded into projective three-space, it is still unknown whether or not every such curve is the intersection of two surfaces.

One of the most active areas of research in algebraic geometry today is the search for a distinguished representative for each birational equivalence class of smooth projective varieties. For curves, we mentioned that every class is represented by a smooth projective curve and that this representative is unique up to isomorphism. In contrast, we cannot find a unique representative for surfaces (two-dimensional varieties). Each birational equivalence class of surfaces contains infinitely many nonisomorphic smooth "models," that is, smooth projective varieties representing the class. For example, it is not very difficult to prove that if we blow up n points on $\mathbb{P}^2$, we obtain a smooth projective variety birationally equivalent to $\mathbb{P}^2$, but for different values of n these varieties are not isomorphic.

A beautiful fact of classical algebraic geometry is that with some describable exceptions, every birational equivalence class of surfaces has a unique *minimal model*. A minimal model is a variety V to which every variety birationally equivalent to V admits a birational, regular (that is, everywhere defined) morphism. Thus the minimal model is a distinguished representative of the birational equivalence class. The exceptions are the birational equivalence classes of *ruled surfaces*—surfaces birationally equivalent to $\mathbb{P}^1 \times C$, where C is a curve. For example $\mathbb{P}^2$ is birationally equivalent to the ruled surface $\mathbb{P}^1 \times \mathbb{P}^1$. Both models are "minimal" in a certain weaker sense: There are no nontrivial birational morphisms from either to any other variety in this birational equivalence class. However, because there are no birational morphisms from $\mathbb{P}^2$ to $\mathbb{P}^1 \times \mathbb{P}^1$, or from $\mathbb{P}^1 \times \mathbb{P}^1$ to $\mathbb{P}^2$, this class does not admit a (unique) minimal model. A fun exposition of this topic can be found in the article by Miles Reid [34].

There is a similar theory for three-folds (algebraic varieties of dimension 3), a beautiful subject for which Mori was awarded the Fields medal in 1990. Again, with certain describable exceptions, each birational equivalence class of three-folds admits a "minimal model," although it is not quite unique and we must allow certain mild singularities, called "terminal singularities." Furthermore, again with certain exceptions, every birational equivalence class contains a "canonical model" to which all other members of the class map, although its singularities are somewhat more complicated.

The higher-dimensional theory is the subject of current research. An enjoyable and very accessible introduction to this topic can be found in Kollár's article [26].

8
Maps to Projective Space

One of the main goals of algebraic geometry is to understand the geometry of smooth projective varieties. For instance, given a smooth projective surface X, we can ask a host of questions whose answers might help illuminate its geometry. What kinds of curves does the surface contain? Is it covered by rational curves, that is, curves birationally equivalent to $\mathbb{P}^1$? If not, how many rational curves does it contain, and how do they intersect each other? Or is it more natural to think of the surface as a family of elliptic curves (genus-1 Riemann surfaces) or as some other family? Is the surface isomorphic to $\mathbb{P}^2$ or some other familiar variety on a dense set? What other surfaces are birationally equivalent to X? What kinds of automorphisms does the surface have? What kinds of continuously varying families of surfaces does it fit into?

In trying to understand a variety X, we might try to make X as concrete as possible. Preferably, we know how to embed X in a particular projective space. It would be even better if we could view X from several different perspectives, by embedding it into projective spaces of different dimensions, and in different ways. More generally, it may be helpful to know whether the variety admits a map to some other projective variety.

Basically, we would like to understand all the maps from a variety to projective space. The main goal of this chapter is to describe how *line bundles* can be used to completely describe these maps. A comprehensive development of this topic would occupy an entire semester in an advanced algebraic geometry course. We hope that our brief treatment here will give the reader an appreciation of this important topic.

Before dealing with line bundles, we first discuss the more elementary topic of embedding a smooth curve in $\mathbb{P}^3$.

8.1   Embedding a Smooth Curve in Three–Space

Suppose that we are given a smooth projective variety. What is the smallest dimensional projective space in which it can be embedded? The following theorem answers this question in the one-dimensional case.

Theorem: Every smooth projective curve can be embedded in $\mathbb{P}^3$.

Before describing the main ideas of the proof we state some definitions.

Definition: Let $X \subset \mathbb{P}^n$ be a smooth projective variety. The *tangent variety* and the *secant variety* of X are defined to be the sets

$$\mathrm{Tan}\, X = \{p \in \mathbb{P}^n \mid p \text{ lies on a line tangent to } X\} \subset \mathbb{P}^n,$$
$$\mathrm{Sec}\, X = \{p \in \mathbb{P}^n \mid p \text{ lies on a line secant to } X\} \subset \mathbb{P}^n,$$

where a line is *secant* to a variety X if it intersects X in at least two distinct points. It is not hard to prove that Tan X and Sec X are quasi-projective varieties (see [20, page 310]). While Tan X is closed for smooth projective X, the variety Sec X is virtually never closed; in fact, tangent lines can be thought of as limits of secant lines, so Tan X is contained in the closure of Sec X.

If X is a smooth curve in $\mathbb{P}^n$, it has only one tangent line at each point x. Let us parametrize the tangent line at x by $\mathbb{P}^1$:

$$\mathbb{P}^1 \xrightarrow{\ \varphi\ } \mathbb{P}^n,$$
$$\lambda \longmapsto \varphi_x(\lambda),$$

where $\phi_x(\lambda)$ is the point on the tangent line at x corresponding to λ under the parametrization ϕ_x.

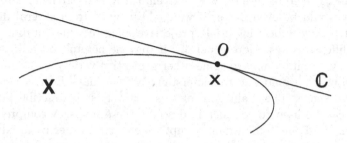

Figure 8.1. Parameterizing the tangent line to a curve

The tangent variety $\text{Tan}\, X$ of a smooth curve X is the image set of the map

$$\begin{aligned} X \times \mathbb{P}^1 &\longrightarrow \mathbb{P}^n, \\ (x, \lambda) &\longmapsto \varphi_x(\lambda). \end{aligned}$$

Since $X \times \mathbb{P}^1$ has dimension two, the image variety $\text{Tan}\, X$ is at most two-dimensional.

Similarly, each secant is determined by two points on X and is parametrized by $\mathbb{P}^1$. This defines a rational map

$$\begin{aligned} X \times X \times \mathbb{P}^1 &\dashrightarrow \mathbb{P}^n, \\ (x, y, \lambda) &\longmapsto \varphi_{x,y}(\lambda), \end{aligned}$$

where $\varphi_{x,y}(\lambda)$ is the point on the secant line between x and y corresponding to $\lambda \in \mathbb{P}^1$ under the parametrization $\varphi_{x,y}$. This map is a morphism on the open set where x and y are distinct, and the image of this morphism is the secant variety. Thus the secant variety has dimension at most three.

Sketch of proof of theorem: Let $X \subset \mathbb{P}^n$ be a smooth curve. Since we want to embed X into projective space $\mathbb{P}^3$, there is no problem at all unless n is at least four. Choose some hyperplane $H \subset \mathbb{P}^n$. Also choose a point $p \in \mathbb{P}^n$, outside both the curve and the hyperplane. Let $\mathbb{P}^n \overset{\pi}{\dashrightarrow} H \cong \mathbb{P}^{n-1}$ be the projection from p to H. Restricted to the curve X, the projection π defines a well-defined morphism $X \to H$, because the only point at which π is not defined is p, and p is not on X. We will show that when n is greater than three, there is a choice of p (in fact, almost all choices work) such that the map $\pi|_X$ is an embedding, completing the proof by induction on n.

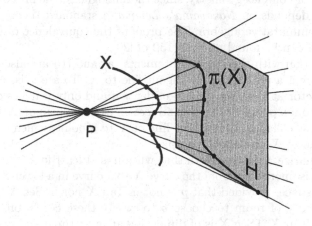

Figure 8.2. Projection of the curve X to H

Evidently, if p does not lie on any secant line to X, then π is one-to-one on X. If $\pi|_X$ is one-to-one, then the following statements are equivalent:

(a) π induces an isomorphism $X \to \pi(X) \subset \mathbb{P}^{n-1}$.

(b) π induces an injection on each tangent space

$$T_x X \xrightarrow{\pi} T_{\pi(x)}(\pi(X)).$$

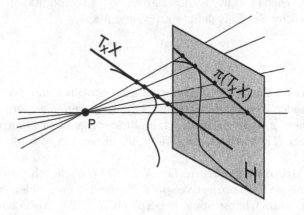

Figure 8.3. Projection of the tangent space

More accurately, the induced map on tangent spaces should be denoted by $d_x\pi$, but in this case $d_x\pi$ is the same as π restricted to the linear variety $T_x X$ in $\mathbb{P}^n$ because π is a projection. The equivalence of (a) and (b) is not completely trivial, although it should be believable if you have studied differential or complex geometry, since the analogue is true in those settings. The proof depends on *Nakayama's Lemma*, a standard theorem in every book on commutative algebra. The proof of the equivalence of statements (a) and (b) can be found on page 152 of [20].

Let us go on with the proof. Statements (a) and (b) are also equivalent to p not being a point on any tangent line to X. To see this, note that a tangent vector at x is sent to zero under π if and only if it lies on the line passing through p and x. Thus, π induces an isomorphism on X if and only if it does not collapse any tangent line of X to a point, which is to say, no tangent line to X passes through p.

To summarize, we have shown that when n is at least four, the projection induces an isomorphism from the curve X to a curve in a lower-dimensional projective space, provided that p is not in $\text{Tan } X$ nor in $\text{Sec } X$. Of course, there is plenty of room to choose p to satisfy these conditions, since the subvariety $\text{Tan } X \cup \text{Sec } X$ is of dimension at most three and can never fill all of $\mathbb{P}^n$ when n is greater than three. $\qquad\square$

Exercise 8.1.1. Use the same techniques to show that every smooth quasi-projective variety of dimension d can be embedded in $\mathbb{P}^{2d+1}$.

8.2 Vector Bundles and Line Bundles

Maps from a variety X to projective space are governed by line bundles on X. A line bundle is the one-dimensional avatar of a *vector bundle*.

Roughly speaking, a vector bundle on X is a morphism of varieties $E \to X$, where E is locally a product of X with $\mathbb{C}^n$, and the map $E \to X$ is locally the natural projection $X \times \mathbb{C}^n \to X$. The precise definition follows.

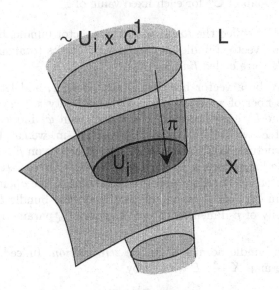

Figure 8.4. A vector bundle L of rank 1

Definition: A *vector bundle of rank n* on an algebraic variety X is an algebraic variety E, together with a morphism $E \xrightarrow{\pi} X$ called the *projection*, such that the following conditions are satisfied:

- There is an open cover $\bigcup U_i$ of X such that $\pi^{-1}(U_i)$ is isomorphic to the product $U_i \times \mathbb{C}^n$ by fiber-preserving maps. More precisely, there are isomorphisms $\pi^{-1}(U_i) \xrightarrow{\varphi_i} U_i \times \mathbb{C}^n$ such that the diagram commutes, where $U_i \times \mathbb{C}^n \xrightarrow{p} U_i$ is the natural projection onto the first factor.

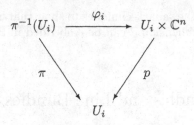

- The isomorphisms φ_i are linearly compatible in the following sense: On $U_i \cap U_j$, the composition

$$\varphi_j \circ \varphi_i^{-1} : (U_i \cap U_j) \times \mathbb{C}^n \longrightarrow (U_i \cap U_j) \times \mathbb{C}^n,$$
$$(x, v) \longmapsto (x, (\varphi_j \circ \varphi_i^{-1})(v)),$$

is a linear map of $\mathbb{C}^n$ for each fixed value of x.

The variety E is called the *total space* of the vector bundle, but we often denote the entire vector bundle by the notation for its total space. Vector bundles of rank 1 are called *line bundles*.

Let $E \xrightarrow{\pi} X$ be a vector bundle of rank n on X, and let $x \in X$ be any point. The fiber of π over x is a closed subvariety $\pi^{-1}(x)$ denoted by E_x. Fixing some U_i containing x, the isomorphism φ_i induces a variety isomorphism of E_x with $\mathbb{C}^n$. Using this isomorphism, we can transfer the vector space structure on $\mathbb{C}^n$ to a vector space structure on E_x. It is easy to verify, using the linear compatability condition (2) above, that this vector space structure on E_x is independent of the choice of the open set U_i and the isomorphism φ_i. Thus, we can think of the vector bundle $E \to X$ as a continuous family of n-dimensional vector spaces E_x parametrized by the points of X.

Every vector bundle admits a unique *zero section*. Indeed, one easily checks that the map $X \xrightarrow{s} E$ defined by

$$x \longmapsto \varphi_i^{-1}(x, 0),$$

where U_i is any open set containing x and 0 is the zero element in $\mathbb{C}^n$, is a well-defined morphism of algebraic varieties. This morphism is called the *zero section* of the bundle because it chooses for us a distinguished point $s(x)$ of each fiber E_x, namely, the zero element of the vector space E_x. Note that $\pi \circ s$ is the identity map on X, so that we can interpret this map as an embedding of X into E. Thus we can think of a vector bundle as a way of continuously attaching an n-dimensional vector space by its origin to each point of X.

The open cover U_i together with the choice of the isomorphisms φ_i is called a *local trivialization* of the bundle. The cover and the maps are by no means unique, and should not be considered as part of the structure of the vector bundle. However, the induced vector space structure on each E_x is unique, which is to say, it is independent of the choice of local trivialization, as the reader should check.

Given a map of algebraic varieties $X \xrightarrow{f} Y$ and a vector bundle $E \xrightarrow{\pi} Y$ on Y, there is a natural way to construct a *pullback bundle* f^*E on X. To get an idea how the pullback is constructed, first recall the notion of a fibered product of sets. If $X \xrightarrow{f} Y$ and $E \xrightarrow{\pi} Y$ are two maps of sets, the fibered product is the set

$$X \times_Y E = \{(x,v) | f(x) = \pi(v)\} \subset X \times E,$$

together with the natural projections $X \times_Y E \xrightarrow{\pi'} X$ and $X \times_Y E \to E$. If $X \xrightarrow{f} Y$ and $E \xrightarrow{\pi} Y$ are both morphisms of algebraic varieties, then the fibered product $X \times_Y E$ has the structure of an algebraic variety, and the projections are morphisms of varieties. If $E \xrightarrow{\pi} Y$ is a vector bundle of rank n over Y, then it can be checked that $X \times_Y E \xrightarrow{\pi'} X$ is a vector bundle, also of rank n, on X. Indeed, for any $x \in X$, the fiber of π' over x is the vector space $\{(x,v) | v \in \pi^{-1}(f(x))\}$, which is identified with the fiber of $E \xrightarrow{\pi} Y$ over the point $f(x)$ and so is isomorphic to $\mathbb{C}^n$. We leave the detailed verification of the vector bundle properties to the reader. The vector bundle $X \times_Y E \to X$ is usually denoted by f^*E.

In particular, if X and Y are isomorphic algebraic varieties, then every line (vector) bundle on X corresponds to a unique line (vector) bundle on Y, its pullback under the isomorphism. Thus the collection of all line bundles on an algebraic variety is an invariant of the variety.

Exercise 8.2.1. If $\{U_i\}$ is a local trivialization of a vector bundle E on Y and $X \xrightarrow{f} Y$ is a morphism of algebraic varieties, show that $\{f^{-1}(U_i)\}$ is a local trivialization of f^*E.

8.3   The Sections of a Vector Bundle

In algebraic geometry it is common to take a different approach to vector bundles, emphasizing the equivalent data of the *sheaf of sections*, rather than the definition above.

Definition: Let $E \xrightarrow{\pi} X$ be a vector bundle, and let $U \subseteq X$ be an open set. A *section* of the vector bundle over the set U is a morphism $U \xrightarrow{s} E$ such that $\pi \circ s$ is the identity map on U. The set of all sections of E over U is denoted by $\mathcal{E}(U)$.

Of course, by our definition of vector bundle, we already know that every vector bundle admits at least one section over each open set, namely the zero section, assigning to each x the zero element in the fiber E_x.

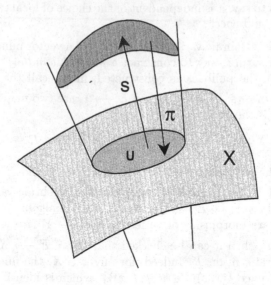

Figure 8.5. Section of a line bundle

If $s_1, s_2 \in \mathcal{E}(U)$ are two sections over U of a vector bundle E on X, then $s_1 + s_2$ is also a section over U. Furthermore, for any regular function $f \in \mathcal{O}_X(U)$, it is easy to show that the product fs is a section of $\mathcal{E}(U)$: The map $U \xrightarrow{fs} \pi^{-1}(U)$ is defined by $x \mapsto f(x) \cdot s(x)$, where $f(x) \in \mathbb{C}$ simply acts by multiplication on the vector $s(x)$ in the vector space $\pi^{-1}(x)$.

The sections of a vector bundle are another example of a sheaf. The reader familiar with the definition of a sheaf (see the Appendix, Section A.1) should have no trouble proving that $\mathcal{E}$ is a *sheaf of modules* over the sheaf of rings $\mathcal{O}_X$: For each open set $U \subset X$, $\mathcal{E}(U)$ is a module over the ring $\mathcal{O}_X(U)$ of regular functions on U. In fact, closer inspection reveals that because E looks locally like $X \times \mathbb{C}^n$, $\mathcal{E}$ is a *locally free* sheaf of $\mathcal{O}_X$-modules of rank n. That is, on sufficiently small open sets,

$$\mathcal{E}(U) \cong \underbrace{\mathcal{O}_X(U) \oplus \cdots \oplus \mathcal{O}_X(U)}_{n \text{ copies}}.$$

Indeed, for such U, a section is a morphism

$$\begin{aligned} U &\longrightarrow \pi^{-1}(U) \cong U \times \mathbb{C}^n, \\ x &\longmapsto (x, f_1(x), \ldots, f_n(x)), \end{aligned}$$

where each of the f_i is a regular function from U to $\mathbb{C}$.

A frequent abuse of notation in algebraic geometry is to use the same symbol to denote both the vector bundle and its sheaf of sections, but we will try to emphasize the difference in our discussion. The sheaf of sections of a vector bundle is one example of what algebraic geometers call a *coherent sheaf* on X. The theory of coherent sheaves, in which one studies more general sheaves of $\mathcal{O}_X$-modules than those that are locally free, is an essential part of algebraic geometry, but we will not go into it here (see [20, Chapter II]).

The *global sections* of a vector bundle are simply the sections $\mathcal{E}(X)$ of E over the whole variety X. Some authors denote the global sections by $\Gamma(X, \mathcal{E})$ or by $H^0(X, \mathcal{E})$. Like the sections over any open set, the set of global sections naturally forms a complex vector space.

It turns out that if X is projective, then the set of global sections $\mathcal{E}(X)$ forms a finite-dimensional $\mathbb{C}$-vector space. The reader familiar with complex geometry will have no trouble believing this, at least when X is smooth, because in this case X is a compact complex manifold. For a precise algebraic proof of much more general facts, see [20, page 122, Theorem 5.19].

Exercise 8.3.1. Let $\mathcal{E}$ be the sheaf of sections of a vector bundle $E \to Y$ and let $X \xrightarrow{f} Y$ be a morphism of algebraic varieties. Describe the sheaf of sections of the pullback bundle f^*E.

8.4   Examples of Vector Bundles

Let $X \subset \mathbb{P}^n$ be a projective variety.

The trivial bundle : The *trivial line bundle* over X is

$$
\begin{aligned}
X \times \mathbb{C} \ & \xrightarrow{\pi} X, \\
(p, \lambda) \ & \longmapsto p.
\end{aligned}
$$

The sections are the morphisms $p \mapsto (p, f(p))$, so giving a section of the trivial bundle over an open set U is the same as giving a regular function $U \xrightarrow{f} \mathbb{C}$. Thus, the sheaf of sections of the trivial line bundle over X can be identified with the structure sheaf $\mathcal{O}_X$ of the variety X. If we assume that X is connected as well as projective, then there are no global sections of the structure sheaf except constants: $\mathcal{O}_X(X) = \mathbb{C}$.

Similarly, the trivial vector bundle on X is the variety $X \times \mathbb{C}^n$, together with the natural projection. Its sheaf of sections is isomorphic to

$$
\underbrace{\mathcal{O}_X \oplus \cdots \oplus \mathcal{O}_X}_{n \text{ copies}}.
$$

The tautological line bundle: All projective varieties embedded in $\mathbb{P}^n$ have a natural line bundle called the *tautological bundle* which is inherited because of the embedding. Indeed, since the points of $\mathbb{P}^n$ are precisely the lines through the origin in $\mathbb{C}^{n+1}$, we can associate to each point $[x_0 : \cdots :$

$x_n] \in \mathbb{P}^n$ the corresponding line $L := \{(tx_0, \ldots, tx_n)|\ t \in \mathbb{C}\}$ in $\mathbb{C}^{n+1}$. By restricting to any subvariety X of $\mathbb{P}^n$, we similarly associate a line to each point of X.

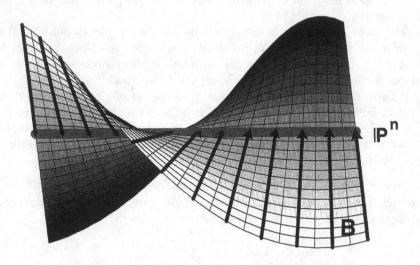

Figure 8.6. The tautological bundle over $\mathbb{P}^1$

More precisely, the tautological bundle over $\mathbb{P}^n$ is constructed as follows. Consider the incidence correspondence of points in $\mathbb{C}^{n+1}$ lying on lines through the origin,

$$B = \{(x, \ell)|\ x \in \ell\} \subseteq \mathbb{C}^{n+1} \times \mathbb{P}^n,$$

together with the natural projection

$$B \xrightarrow{\pi} \mathbb{P}^n.$$

(The same set B, taken with the other projection $B \to \mathbb{A}^{n+1}$, defines the blowup of the origin in $\mathbb{A}^{n+1}$.) The fiber over a fixed point $\ell \in \mathbb{P}^n$ is the set $\{(x, \ell)|\ x \in \ell\}$ of points x on ℓ. It is easy to verify that the projection $B \xrightarrow{\pi} \mathbb{P}^n$ satisfies the definition of a line bundle.

The tautological bundle over the projective variety $X \subseteq \mathbb{P}^n$ is obtained by simply restricting the correspondence to the points of X:

$$B = \{(x, \ell)|\ \ell \in X \text{ and } x \in \ell\} \subseteq \mathbb{C}^{n+1} \times X \subset \mathbb{C}^{n+1} \times \mathbb{P}^n.$$

It is important to realize that the tautological bundle is not intrinsic to X: It depends on the choice of the embedding of X in a particular projective space. In other words, the pullback of the tautological bundle under an isomorphism may fail to be tautological for a different embedding. See the exercise at the end of this section for an example of this phenomenon.

Tautological line bundles have no nonzero global sections at all. To understand why this is true, it helps to consider first the case of the projective line. A global section of the tautological bundle of $\mathbb{P}^1$ defines, for each point p in $\mathbb{P}^1$, a point $(a(p), b(p)) \in \mathbb{C}^2$ lying on the line through the origin corresponding to p. Since the assignment

$$\mathbb{P}^1 \longrightarrow \mathbb{C}^2,$$
$$p \longmapsto (a(p), b(p)),$$

must be a morphism of algebraic varieties, we see that projecting onto either factor, we have morphisms (regular functions),

$$\mathbb{P}^1 \longrightarrow \mathbb{C}$$
$$p \longmapsto a(p)$$

and

$$p \longmapsto b(p).$$

But because $\mathbb{P}^1$ admits no non-constant regular functions, both regular functions a and b are constant functions. But then both are zero functions: The point $(a(p), b(p))$ is supposed to lie on the line in $\mathbb{C}^2$ corresponding to p, and the only point of $\mathbb{C}^2$ that lies on all lines through the origin is $(0, 0)$ itself. Thus the tautological bundle over $\mathbb{P}^1$ admits only the zero section. The obvious extension of this argument shows that the tautological line bundle over any projective variety (which is not just a finite collection of points) has no nonzero global sections.

The sheaf of sections of the tautological line bundle over X is often denoted by $\mathcal{O}_X(-1)$. Implicit in such notation is the existence of a particular embedding of X in $\mathbb{P}^n$.

The hyperplane bundle: The *hyperplane* bundle H on a quasi-projective variety is defined to be the *dual* of the tautological line bundle: The fiber $\pi^{-1}(p)$ over a point $p \in X \subset \mathbb{P}^n$ is the (one-dimensional) vector space of linear functionals on the line $\ell \subset \mathbb{C}^{n+1}$ that determines p in $\mathbb{P}^n$. The formal construction of H as a subvariety of $(\mathbb{C}^{n+1})^* \times \mathbb{P}^n$ parallels that of the tautological bundle.

This line bundle has many global sections. Indeed, let $\sum_{i=0}^n a_i x_i$ be any linear functional on $\mathbb{C}^{n+1}$. For a point $p = [\lambda_0 : \lambda_1 : \ldots : \lambda_n] \in X$, this linear form can be restricted to the line $\ell = \{(t\lambda_0, t\lambda_1, \ldots, t\lambda_n) \mid t \in \mathbb{C}\} \subset \mathbb{C}^{n+1}$ corresponding to p. This gives a well-defined global section

$$X \longrightarrow H,$$
$$p \longmapsto (p, \sum_{i=0}^n a_i x_i |_\ell),$$

of the hyperplane bundle. In fact, one can check that the global sections of the hyperplane bundle on $\mathbb{P}^n$ are precisely the linear polynomials in $\mathbb{C}[x_0, \ldots, x_n]$.

The sheaf of sections of the hyperplane bundle on a subvariety X of $\mathbb{P}^n$ is usually denoted by $\mathcal{O}_X(1)$. Again, implicit in this notation is the existence of a specific embedding of X in a projective space $\mathbb{P}^n$. For a different embedding, a different line bundle may become the hyperplane bundle.

Square of the hyperplane bundle: The *square H^2 of the hyperplane bundle* on a projective subvariety X of $\mathbb{P}^n$ associates to each $p \in X$ the vector space of all quadratic homogeneous polynomials on the line $\ell \subset \mathbb{C}^{n+1}$ determining $p \in \mathbb{P}^n$. The sheaf of sections of the square of the hyperplane bundle is denoted by $\mathcal{O}_X(2)$. For example, the global sections of $\mathcal{O}_{\mathbb{P}^n}(2)$ are precisely the homogeneous polynomials of degree 2 in $\mathbb{C}[x_0, \ldots, x_n]$. Like the hyperplane and tautological bundles, H^2 depends on a particular embedding of X in $\mathbb{P}^n$.

Duals and products in general: All the standard linear algebra constructions for vector spaces are also valid for vector bundles. For example, if $E \to X$ is a vector bundle with fiber over p denoted by E_p, then there exist vector bundles

$$E^* \to X$$

whose fibers are the dual spaces $(E_p)^*$, and

$$\overset{i}{\bigwedge} E \to X$$

whose fibers are the exterior products $(\bigwedge^i E_p)$. If $F \to X$ is another vector bundle over X, then there exists a vector bundle

$$E \otimes F \longrightarrow X$$

whose fibers are $(E_p \otimes F_p)$, and

$$E \oplus F \to X$$

whose fibers are $E_p \oplus F_p$. We leave the careful construction of these bundles as an instructive exercise for the reader.

The line bundle H^2 above is the tensor product $H \otimes H$, where H is the hyperplane bundle on X. More generally, one can define the rth power $E^{\otimes r}$ of any vector bundle E to be the r-fold tensor product of E with itself. On $\mathbb{P}^n$, the rth power of the hyperplane bundle, whose sheaf of sections is denoted by $\mathcal{O}_{\mathbb{P}^n}(r)$, has fibers described as follows: The fiber over a point $\ell \in \mathbb{P}^n$ consists of all degree r homogeneous polynomials on the one-dimensional vector space corresponding to ℓ in $\mathbb{C}^{n+1}$. Each homogeneous polynomial of degree-r in $\mathbb{C}[x_0, \ldots, x_n]$ determines such a function when restricted to any line in $\mathbb{C}^{n+1}$ and therefore gives rise to a global section. In fact, the space of all global sections of $\mathcal{O}_{\mathbb{P}^n}(r)$ can be identified with the space of all homogeneous degree r polynomials in $\mathbb{C}[x_0, \ldots, x_n]$.

We caution the reader that we write $\mathcal{O}_{\mathbb{P}^n}(r)$ for the r-fold tensor product of $\mathcal{O}_{\mathbb{P}^n}(1)$ with itself, but algebraic geometers do not mean the r-fold product of E when they write $E(r)$ for some vector (or line) bundle E. Rather, the notation $E(r)$ means $E \otimes \mathcal{O}_{\mathbb{P}^n}(r)$.

The tangent bundle and its relatives: Let X be an irreducible smooth quasi-projective variety of dimension n. Associated to X are several natural vector bundles and line bundles arising from the tangent space.

The *tangent bundle* is a rank-n vector bundle $TX \to X$ such that the fiber over any point $p \in X$ is the tangent vector space $T_p X$. The total space of the tangent bundle was introduced in Section 6.2. The *cotangent bundle* $T^* X \to X$ is its dual: The fiber over any point $p \in X$ is the cotangent space $(T_p X)^*$. The sections of the cotangent bundle are called *differential one-forms*. The sheaf of sections of the tangent bundle is often denoted by Θ_X, while Ω_X frequently denotes the sheaf of sections of the cotangent bundle.

Unlike the tautological and hyperplane bundles, the tangent bundle and its dual are independent of the embedding of X into projective space, although this is not obvious from our description of it. In other words, if $X \xrightarrow{f} Y$ is an isomorphism, then the tangent bundle on Y pulls back to the tangent bundle on X, and similarly for the cotangent bundle. Thus the tangent and cotangent bundles are intrinsically defined objects attached to X.

If we fix a local trivialization of the tangent bundle, so that over an open set $U \subset X$ the tangent bundle is isomorphic to $U \times \mathbb{C}^n$, then we can denote the "linear functionals"

$$
\begin{aligned}
\{p\} \times \mathbb{C}^n &\longrightarrow \mathbb{C}, \\
(p, (\lambda_1, \ldots, \lambda_n)) &\longmapsto \lambda_i,
\end{aligned}
$$

by $d_p x_i$. As we vary over all p in U, we get a section dx_i of the cotangent bundle over U because $U \xrightarrow{dx_i} U \times \mathbb{C}^{n*}$ is a regular map and for each point $p \in U$ it defines a linear functional $T_p X = \mathbb{C}^n \xrightarrow{d_p x_i} \mathbb{C}$. Because they form a basis for $(T_p X)^*$, every section of the cotangent bundle can be written locally on U as

$$
f_1 \, dx_1 + \cdots + f_n \, dx_n,
$$

where the $f_i \in \mathcal{O}_X(U)$ are regular functions and dx_i acts on a vector λ in the tangent space at p by $d_p(x_i)(\lambda)$. In algebraic, differential, and complex geometry, differential forms have similar local descriptions. In algebraic geometry the f_i are regular (polynomial) functions, in complex geometry the f_i are holomorphic, and in differential geometry the f_i are smooth functions.

The canonical line bundle: The most frequently encountered line bundle in algebraic geometry (with the exception of the trivial bundle $\mathcal{O}_X$) is the *canonical bundle*. If X is a smooth irreducible variety of dimension n, then the canonical line bundle is the highest exterior power of its cotangent bundle

$$\bigwedge^n T^*X.$$

The sheaf of sections of the canonical bundle is denoted by ω_X. As in the previous example, its sections over a sufficiently small open set U can be written as

$$f dx_1 \wedge \cdots \wedge dx_n,$$

where f is a regular function on U. The canonical bundle is important because it and its powers are the only line bundles on an algebraic variety that are intrisically defined.

Exercise 8.4.1. Let $\mathbb{P}^1 \xrightarrow{\nu_n} \mathbb{P}^n$ be the Veronese embedding of $\mathbb{P}^1$ as a rational normal curve C_n of degree n in $\mathbb{P}^n$. Prove that the tautological bundle on C_n pulls back under ν_n to the nth power of the tautological bundle on $\mathbb{P}^1$. This example shows that the "tautological bundle" depends on the embedding in projective space.

8.5   Line Bundles and Rational Maps

Our next goal is to illustrate how line bundles and their global sections govern all rational maps of varieties to $\mathbb{P}^n$. An understanding of all the possible ways in which a variety may be mapped to projective spaces is tantamount to a complete understanding of all line bundles on the variety.

Let X be a quasi-projective variety and let $L \xrightarrow{\pi} X$ be a line bundle over X. Let us choose a set $\{s_0, \ldots, s_n\}$ of linearly independent sections of the $\mathbb{C}$-vector space of its global sections. The vector space spanned by these sections is called a *linear system* on X; if this vector space consists of all the global sections of L, it is called a *complete linear system*. A complete linear system is often denoted by $|L|$. Using these sections, we define the rational map

$$\begin{aligned} X &\dashrightarrow \mathbb{P}^n, \\ x &\longmapsto [s_0(x) : \ldots : s_n(x)]. \end{aligned}$$

We consider an example before clarifying the meaning of this map.

Example: Consider the hyperplane bundle H on $\mathbb{P}^n$. A basis for its global sections is $x_0, x_1, \ldots, x_n$, where the x_i are the homogeneous coordinates of $\mathbb{P}^n$. Consider the (incomplete) linear system spanned by $x_0, \ldots, x_{n-1}$. The associated rational map is

$$\begin{aligned} \mathbb{P}^n &\dashrightarrow \mathbb{P}^{n-1}, \\ [x_0 : \cdots : x_n] &\longmapsto [x_0 : \cdots : x_{n-1}]. \end{aligned}$$

This map is defined everywhere except at the point $p = [0 : 0 : \cdots : 0 : 1]$. As we have seen before, the map is simply projection from the point p onto the hyperplane $\mathbb{V}(x_n) \cong \mathbb{P}^{n-1}$ in $\mathbb{P}^n$.

We now discuss in detail the meaning of the expression $[s_0 : \cdots : s_n]$ where s_i are sections of a line bundle. First of all, the sections s_i are not functions, so the meaning of $s_i(x)$ must be interpreted so as to make the $(n+1)$-tuple $[s_0(x) : \cdots : s_n(x)]$ an actual point in $\mathbb{P}^n$. Choosing a local trivialization for $L \xrightarrow{\pi} X$, we have an identification of $\pi^{-1}(U) \subseteq L$ with $U \times \mathbb{C}$ in a neighborhood U of x. This allows us to identify the section

$$U \xrightarrow{s_i} \pi^{-1}(U) \xrightarrow{\cong} U \times \mathbb{C},$$
$$x \longmapsto s_i(x) \longmapsto (x, \tilde{s}_i(x)),$$

with the regular function $U \to \mathbb{C}$ sending x to $\tilde{s}_i(x)$. When we write the $(n+1)$-tuple $[s_0(x) : \cdots : s_n(x)]$, we really mean the $(n+1)$-tuple of complex numbers $[\tilde{s}_0(x) : \cdots : \tilde{s}_n(x)]$. Now, how do we know that this $n + 1$ tuple does not depend on our choice of local trivialization? We do not know this, and in fact, it is not true: A different choice of a trivialization produces a different vector. But because the local trivializations of a line bundle are compatible with linear changes of coordinates, it is easy to check that *up to nonzero scalar multiple*, the vector $[s_0(x) : \cdots : s_n(x)]$ is well-defined. That is, if one local trivialization produces the $(n+1)$-tuple $[\tilde{s}_0(x) : \cdots : \tilde{s}_n(x)]$ of complex numbers and another produces $[\tilde{\tilde{s}}_0(x) : \cdots : \tilde{\tilde{s}}_n(x)]$, then there exists a regular function λ in a neighborhood of x such that $\lambda(x)$ is a nonzero complex number and for all i,

$$\tilde{s}_i(x) = \lambda(x)\, \tilde{\tilde{s}}_i(x).$$

That is,

$$[\tilde{s}_0 : \cdots : \tilde{s}_n] = [\tilde{\tilde{s}}_0 : \cdots : \tilde{\tilde{s}}_n],$$

and the notation $[s_0(x) : \cdots : s_n(x)]$ represents a well-defined element of $\mathbb{P}^n$.

The only remaining problem occurs when the sections s_i all vanish at x, so that $[s_0(x) : \cdots : s_n(x)]$ is the zero $(n+1)$-tuple. Unfortunately, there is nothing to prevent the s_i from simultaneously vanishing, which is why the map

$$X \dashrightarrow \mathbb{P}^n,$$
$$x \longmapsto [s_0(x) : \cdots : s_n(x)],$$

is only a rational map and not an everywhere defined morphism of varieties. The rational map is defined on the open set in X complementary to the common vanishing set of the sections s_i.

The rational map $X \dashrightarrow \mathbb{P}^n$ depends on the choice of the basis $\{s_0, \ldots, s_n\}$ for the linear system, but the reader will quickly verify that different bases produce maps that can be transformed to each other by an automorphism of $\mathbb{P}^n$.

The construction can be reversed: Every rational map $X \dashrightarrow \mathbb{P}^n$ is determined by some linear system of some line bundle over X. Indeed, the line bundle on X will be the pullback of the hyperplane bundle on $\mathbb{P}^n$, and the s_i will be the pullbacks of the coordinate functionals x_i on $\mathbb{P}^n$. (Strictly speaking, the line bundle may be defined only on a dense open subset of X where the rational map is regular, since pulling back makes sense only under a regular map. However, a line bundle defined on a dense open set of X often extends uniquely to the whole of X; this is always the case, for example, when X is smooth.)

It is easy to verify that the common zero set of a set of global sections $\{s_i\}$ of a line bundle is a closed subvariety of X, called the *base locus* of the linear system spanned by the s_i. For example, the base locus of the projection discussed in the previous example is the single point $\{p\}$. If we are dealing with a complete linear system, we call this zero set the base locus of the line bundle L. In the best possible scenario, the base locus is empty, and the resulting rational map is a morphism. Such a linear system is called *base-point free*, and the associated line bundle is said to be *globally generated*. Because globally generated line bundles determine morphisms to projective space, the identification of globally generated line bundles is an important area of research.

Example: Consider the nth power of the hyperplane bundle on $\mathbb{P}^1$, $H^n \xrightarrow{\pi} \mathbb{P}^1$. The fiber over the point $x = [\lambda_0 : \lambda_1] \in \mathbb{P}^1$ consists of all degree-n homogeneous polynomials on the line $\ell = \{(\lambda_0 t, \lambda_1 t) \mid t \in \mathbb{C}\}$ in $\mathbb{C}^2$. To see this, first note that the sheaf of sections $\mathcal{O}_{\mathbb{P}^1}(n)$ is just the n-fold tensor product $\mathcal{O}_{\mathbb{P}^1}(1) \otimes \ldots \otimes \mathcal{O}_{\mathbb{P}^1}(1)$, where $\mathcal{O}_{\mathbb{P}^1}(1)$ is the sheaf of sections of the hyperplane bundle. Recall that the global sections of $\mathcal{O}_{\mathbb{P}^1}(1)$ are spanned by the basis $\{x_0, x_1\}$, that is, they are the linear functionals $a_0 x_0 + a_1 x_1$. Likewise, the global sections of $\mathcal{O}_{\mathbb{P}^1}(n)$ are the n-fold tensor products of these, which are all degree-n homogeneous polynomials of two variables. So the monomials $x_0^n, x_0^{n-1} x_1, \ldots, x_1^n$ form a basis for the space of global sections of the nth power of the hyperplane bundle. For example, if $n = 2$, the corresponding rational map is

$$\mathbb{P}^1 \quad \dashrightarrow \quad \mathbb{P}^2,$$
$$[x_0 : x_1] \quad \longmapsto \quad [x_0^2 : x_0 x_1 : x_1^2],$$

the second Veronese embedding of $\mathbb{P}^1$ in $\mathbb{P}^2$. In general, the rational map given by the complete linear system of the nth power of the hyperplane bundle on an arbitrary variety X is simply the Veronese map ν_n. Because the sections x_0^n and x_1^n do not simultaneously vanish, the line bundle H^n is globally generated, and as we have seen before, the rational map it defines is an everywhere defined morphism of algebraic varieties, the Veronese map.

Example: The embedding of a smooth projective curve into projective three-space constructed in Section 8.1 can be understood in this context. Let X be a smooth curve in $\mathbb{P}^n$, where $n \geq 4$, and let p be a point of $\mathbb{P}^n$ not on the tangent or secant varieties of X. Choosing any set of n linear functionals, $s_1, \ldots, s_n$, vanishing simultaneously precisely at p, we have a rational map

$$\mathbb{P}^n \dashrightarrow \mathbb{P}^{n-1},$$
$$x \longmapsto [s_1(x) : \cdots : s_n(x)],$$

corresponding to the linear system $\{s_1, \ldots, s_n\}$ of the hyperplane bundle on $\mathbb{P}^n$. Restricting to X, we have a linear system that is base-point free on X (as $p \notin X$), and hence determines a morphism

$$X \longrightarrow \mathbb{P}^{n-1},$$
$$x \longmapsto [s_1(x) : \cdots : s_n(x)].$$

After iterating this procedure, we wind up with a rational map $\mathbb{P}^n \dashrightarrow \mathbb{P}^3$ given by a linear system spanned by four global sections s_0, s_1, s_2, s_3 of the hyperplane bundle on $\mathbb{P}^n$, which becomes base-point-free after restriction to X. Thus the corresponding restriction map $X \to \mathbb{P}^3$ is a morphism.

The previous discussion indicates the importance of the zero sets of sections of line bundles.

Definition: The zero set of a nonzero global section of a line bundle is called a *divisor* of the line bundle.

By fixing a local trivialization, a section of a line bundle can be locally identified with a regular function on the variety. Thus a divisor of a line bundle is defined locally by a single equation, that is, it is *locally principal,* and so it has codimension one in the ambient variety. Of course, two different nonzero global sections of a line bundle will usually have different zero sets, so the divisor of a line bundle is not unique.

Examples: Consider the hyperplane bundle on $\mathbb{P}^n$. Its global sections are the linear forms $a_0 x_0 + \ldots + a_n x_n$. Thus its divisors are the zero sets of these linear forms, the hyperplanes, in $\mathbb{P}^n$. The collection of all divisors of the hyperplane bundle in $\mathbb{P}^n$ is the same as the collection of all hyperplanes in $\mathbb{P}^n$.

Consider the square of the hyperplane bundle on $\mathbb{P}^n$. Its global sections are the quadratic forms $\sum_{ij} a_{ij} x_i x_j$. Thus its divisors are the zero sets of these forms, the quadric hypersurfaces in $\mathbb{P}^n$. Note, for example, that the divisor associated to the section x_0^2 should be considered as the "double hyperplane" $2H$, where H is the hyperplane defined by $x_0 = 0$.

As this example indicates, a divisor should really be considered as a zero set with multiplicities, which we can represent as a formal linear combination of irreducible codimension-one subvarieties with integer coefficients. For example, consider the fourth power of the hyperplane bundle on $\mathbb{P}^n$, whose global sections are the homogeneous quartic (degree 4) forms in $n+1$ variables. The divisors associated to this are the irreducible hypersurfaces in $\mathbb{P}^n$ of degree 4, but also divisors defined as the zero sets of, say, $F_3 x_0$, where F_3 is an irreducible cubic. The zero set of $F_3 x_0$ is the union of the irreducible cubic hypersurface C defined by F_3 and the hyperplane H defined by x_0. This divisor can be written in additive notation as $C + H$. Similarly, the zero set of the section $x_0^2 x_1^2$ can be written as $2H + 2H'$, where H (respectively H') is the hyperplane defined by x_0 (respectively x_1).

It is also possible to define divisors associated to line bundles that do not have any global sections by considering the "zeros and poles of a rational section." For example, because the tautological bundle on $\mathbb{P}^n$ is dual to the hyperplane bundle, it stands to reason that its divisors ought to be of the form $-H$, where H is a hyperplane in $\mathbb{P}^n$. These are called "virtual divisors" in the older literature.

Given any divisor of a line bundle, it is possible to reconstruct the line bundle up to isomorphism. Two divisors are *linearly equivalent* if they are associated to the same line bundle, and the equivalence class of all divisors associated to a fixed line bundle is called the *divisor class* or the *Chern class* of the line bundle. Thus, the theory of line bundles (up to isomorphism) is equivalent to the theory of divisors (up to linear equivalence). The Chern class of the canonical bundle is especially important and is called the *canonical class* of the variety.

Although a complete understanding of line bundles requires a thorough understanding of divisors, we do not continue with this important topic here. See [37, Chapter III, Section 1] or [20, Chapter II, Sections 6 and 7] for the basic theory of divisors and line bundles.

Exercise 8.5.1. Let $X \subset \mathbb{P}^2$ be a smooth projective cubic curve, and let H denote the hyperplane bundle on X. Show that every set of three collinear points on X determines a divisor associated to H. To what extent is the converse true?

Exercise 8.5.2. Assume that the smooth projective cubic curve $X \subset \mathbb{P}^2$ is given by an equation of the form $zy^2 = f(z, x)$, where f is a homogeneous degree-three polynomial in x and z. Show that there is a morphism $X \xrightarrow{f} \mathbb{P}^1$ given by $[x : y : z] \longmapsto [x : z]$, and that this is a two-to-one cover of $\mathbb{P}^1$ except at three points of $\mathbb{P}^1$ (called ramification points). What is the linear system determining this map?

Exercise 8.5.3. Let $X \subset \mathbb{P}^n$ be an irreducible projective variety. Describe the divisors associated to the hyperplane bundle on X.

8.6   Very Ample Line Bundles

Line bundles determine rational maps to projective space. The very ample line bundles determine *embeddings* in projective space. Let X be any projective variety.

Definition: A line bundle $L \to X$ is called *very ample* if the rational map determined by its complete linear system $|L|$,

$$X \dashrightarrow \mathbb{P}^n,$$

is an everywhere defined morphism that defines an isomorphism onto its image.

Let $X \to \mathbb{P}^n$ be the embedding of X in projective space determined by a basis $s_0, \ldots, s_n$ of the global sections of a line bundle L on X. Under this morphism, the sections s_i become the coordinate functions x_i. Thus, after embedding X in $\mathbb{P}^n$ this way, the line bundle L has become the hyperplane bundle on $X \subset \mathbb{P}^n$. So we may think of a very ample line bundle as one that, for some embedding of X in projective space is the hyperplane bundle on X. The term "very ample" suggests that a line bundle has very many global sections. Recall that even to make the map given by $|L|$ an everywhere defined morphism, the line bundle L must be globally generated— it must admit enough global sections so that for each point of X, there is some global section of L that does not vanish there. But even in this case the morphism is usually not an isomorphism onto its image, nor even one-to-one. To make the map given by $|L|$ one-to-one, we need even more sections: For any two points of X, there should be a global section of L that vanishes at one but not the other (L must "seperate points"). But a very ample line bundle requires still more global sections, since not every one-to-one morphism is an embedding; see the Example in Section 2.5. To be very ample, a line bundle must "separate tangent vectors" as well.

Example: The positive powers of the hyperplane bundle on a projective variety are very ample line bundles. Indeed, the maps they determine are the Veronese maps, which we proved to be embeddings in Section 5.1.

It follows from the preceding example that if L is a very ample line bundle on X, then every positive power of L is also very ample. However, there are non-very-ample line bundles L with the property that some power L^n is very ample; we will construct an example in the exercises. A line bundle with the property that some positive power is very ample is said to be *ample*.

An ample line bundle L has the following important property: Given any line bundle M, the bundle $M \otimes L^n$ is very ample for all sufficiently large n. An active area of research today is the investigation of how large is "sufficiently large" in this context, especially for the so-called *adjoint bundles* $\omega \otimes L^n$, which play a prominent role in the classification problems. In particular, it is most interesting to find a uniform N, depending only on X, that works for all ample line bundles L.

Open Problem: Let V be a smooth projective variety with canonical bundle ω. Is there a uniform N such that for all $n > N$, the line bundle $\omega \otimes L^n$ is very ample, where L is any ample line bundle on V? What is the best possible such N?

Fujita's conjecture predicts that the best possible value for N is $\dim V + 2$ in general. For curves, the conjecture follows from the Riemann–Roch theorem, and for complex surfaces, it has been proved by Reider. There is little progress on the three–fold case. This and related questions, such as finding bounds such that $\omega \otimes L^n$ is globally generated, occupy a large number of researchers today. See [29] for an overview of recent progress.

The rational map associated with the canonical line bundle is known as the *canonical map*. The canonical bundle and its powers are the only intrinsic line bundles on an algebraic variety, and so they provide the only intrinsic maps to projective space. A good way to compare two varieties is to map each to projective space using the canonical maps and then compare the images. It is especially useful to know whether the canonical line bundle—or some specific power of the canonical line bundle—is very ample. For example, this is helpful in classifying algebraic curves.

Let X be a smooth projective curve, and consider the canonical line bundle ω_X of X, consisting of differential one-forms on X. *Hodge theory* tells us that the dimension of the space of global sections of the canonical line bundle ω_X is

$$\dim(\omega_X(X)) = \text{genus}(X) = g,$$

where g is the topological genus of X considered as a compact Riemann surface (that is, the dimension of $H^1(X, \mathbb{Q})$). Thus the canonical map takes the form

$$X \dashrightarrow \mathbb{P}^{g-1}.$$

For genus-zero curves, we see that ω_X has no nonzero global sections at all, so the canonical map is not defined.

For genus-one curves, the space of global sections of ω_X has dimension one, so the canonical map $X \to \mathbb{P}^0$ simply collapses X to a point.

For genus-two and higher curves, the canonical map is more interesting. It is a fact that on a curve X of genus $g \geq 2$, the canonical bundle is always globally generated (see [20, page 341]), so the canonical map

$$X \to \mathbb{P}^{g-1}$$

is an everywhere-defined morphism of algebraic varieties. Now, there are two possibilities. The canonical bundle is either

- very ample, in which case the canonical map $X \xrightarrow{|\omega|} \mathbb{P}^{g-1}$ is an isomorphism onto its image set, or

- not very ample, in which case $X \to \mathbb{P}^{g-1}$ is not an embedding. However, this is an exceptional case. If this happens, the image set will be isomorphic to $\mathbb{P}^1$ and the map will be generically 2-to-1, meaning that with a finite number of exceptions, or "ramification points," the map is 2-to-1. In this case, the curve X is said to be *hyperelliptic*.

Every curve of genus 2 is hyperelliptic, since the canonical map $X \to \mathbb{P}^1$ is obviously not an embedding. More generally, the hyperelliptic curves form a $(2g-1)$-dimensional subvariety in the moduli space $\mathfrak{M}_g$ of all smooth projective curves of genus g, which is of dimension $3g-3$. Thus a generic curve of genus greater than two will be nonhyperelliptic, and so its canonical bundle is very ample. In this case, the curve admits a *canonical embedding* into projective space such that the hyperplane bundle is the canonical bundle.

Because it is not always possible to canonically embed a curve, we seek other natural embeddings in projective space. It turns out that for any curve of genus greater than two, the square of the canonical bundle is always very ample. Thus we have a completely intrinsic way to embed any such abstract curve in projective space. More generally, one can look at *pluricanonical* maps, that is, maps induced by the complete linear system associated to a power of the canonical bundle.

The intrinsic nature of the pluricanonical maps gives rise to the following important property. Let X be a k-canonically embedded variety in $\mathbb{P}^n$. This means that the inclusion $X \subset \mathbb{P}^n$ is induced by the complete linear system of the kth power of the canonical bundle, or, put differently, that the hyperplane bundle on X is the same as $\omega^{\otimes k}$. Now, if Y is another k-canonically embedded variety in $\mathbb{P}^n$, then X and Y are isomorphic if and only if they are projectively equivalent. This is not difficult to prove, provided that one observes that any isomorphism of curves must preserve the vector space of differential k-forms. So any isomorphism $X \to Y$ is actually a projective equivalence, that is, a mere change of coordinates in the ambient projective space $\mathbb{P}^n$.

From this point of view, one gets an idea of how Mumford constructed his famous moduli spaces. For example, fix a genus $g \geq 3$, and let us sketch the idea of the construction of $\mathfrak{M}_g$. In this case, the square of the canonical bundle of any smooth curve of genus g is very ample. The classical Riemann–Roch theorem is a formula to compute the space of global sections of any line bundle on any smooth curve (see [20, IV, Section 1]); it can be used here to show that the dimension of the space of global sections of the square of the canonical bundle is $3g-3$. Thus, every smooth curve of genus $g \geq 3$ embeds in $\mathbb{P}^{3g-4}$ via the so-called bi-canonical map.

Now, again using the Riemann–Roch theorem, one computes that the Hilbert polynomial of such a curve in $\mathbb{P}^{3g-4}$ is $P(n) = (4g-4)n - g + 1$.

Thus every smooth curve of genus g will be represented by a point of the Hilbert scheme of smooth curves in $\mathbb{P}^{3g-4}$ with Hilbert polynomial P. Conversely, it can be checked that every smooth curve in $\mathbb{P}^{3g-4}$ with Hilbert polynomial P is a curve of genus g. Furthermore, because the curves are bi-canonically embedded, any two such curves are isomorphic if and only if they are projectively equivalent. The group $\mathbf{PGL}(3g-3)$ of automorphisms of $\mathbb{P}^{3g-4}$ acts on the Hilbert scheme, taking a point representing a curve X to the projectively equivalent (isomorphic) curves. In other words, the isomorphism classes of smooth curves of genus g can be interpreted as the orbits of the natural $\mathbf{PGL}(3g-3)$ action on the Hilbert scheme of bicanonically embedded smooth curves of genus g. Put differently, the quotient of the Hilbert scheme by the action of $\mathbf{PGL}(3g-3)$ ought to be a parameter space for the set of smooth curves of genus g. The only remaining difficulty is to equip this quotient with the structure of an algebraic variety. In order to solve this difficult problem, Mumford developed a method for defining quotients in algebraic geometry, which he called *geometric invariant theory* (or "GIT"), and applied it to construct the moduli spaces $\mathfrak{M}_g$ [16]. It turns out that in order to effectively carry out this program, it is not sufficient to consider bi–canonically embedded curves—we must look at k-canonically embedded curves for some very large k—but the same ideas apply.

This is only a very small taste of the rich ideas of how line bundles, and especially the canonical line bundle, can be useful in understanding algebraic varieties. Lazarsfeld's monograph [30] is a good place to begin further reading on the topic of linear systems, whereas Harris's recent book gives a more complete account of the construction of Hilbert schemes and moduli spaces [18]. An elementary account of the Riemann–Roch theorem for curves can be found in Fulton's book [14]. Like the rest of this book, our exposition is merely intended as an advertisement for the deep and beautiful mathematics of algebraic geometry.

Exercise 8.6.1. Prove that k-canonically embedded curves are isomorphic if and only if they differ by a projective automorphism (that is, a linear change of coordinates). Hint: Any isomorphism between curves must take the canonical bundle to the canonical bundle.

Appendix A
Sheaves and Abstract Algebraic Varieties

A.1 Sheaves

Let X be a topological space. For each open set $U \subset X$, consider the set $\mathcal{F}(U, \mathbb{C})$ of all $\mathbb{C}$-valued functions on U. This set naturally forms a $\mathbb{C}$-algebra under pointwise addition and multiplication of functions.

Definition: A *sheaf* $\mathcal{R}$ *of* $\mathbb{C}$-*valued functions* on X assigns to each open set $U \subset X$ a subalgebra $\mathcal{R}(U) \subset \mathcal{F}(U, \mathbb{C})$ in a way that is "compatible with both restriction and gluing," that is,

- For any open sets $U_1 \subset U_2 \subset X$ and $f \in \mathcal{R}(U_2)$, the restriction of f to U_1 is in $\mathcal{R}(U_1)$.

- If $\{U_\alpha\}_{\alpha \in A}$ is an open cover of an open set $U \subseteq X$ and $f \in \mathcal{F}(U, \mathbb{C})$ is such that $f|_{U_\alpha} \in \mathcal{R}(U_\alpha)$ for all α, then $f \in \mathcal{R}(U)$.

The functions $f \in \mathcal{R}(U)$ are called the *sections* of the sheaf $\mathcal{R}$ over the open set $U \subset X$.

If $\mathcal{R}$ is a sheaf of $\mathbb{C}$-valued functions and if $f \in \mathcal{R}(U_1 \cup U_2)$, then note that $f|_{U_1}$ and $f|_{U_2}$ both have the same restriction to $U_1 \cap U_2$, namely $f|_{U_1 \cap U_2}$. Conversely, if $h \in R(U_1)$ and $g \in R(U_2)$ are such that $h|_{U_1 \cap U_2} = g|_{U_1 \cap U_2}$, then the map $f \in \mathcal{F}(U_1 \cup U_2, \mathbb{C})$ given by

$$f(x) = \begin{cases} h(x) \text{ if } x \in U_1, \\ g(x) \text{ if } x \in U_2, \end{cases}$$

is well-defined. Clearly, $f|_{U_1} = h$ and $f|_{U_2} = g$. So by the second property of sheaves, $f \in \mathcal{R}(U_1 \cup U_2)$. We say that h and g glue together to give f.

In a similar way, we can define a *sheaf of* $\mathbb{R}$-*valued functions*, or a *sheaf of functions* with values in any ring, or even in any set.

Examples of sheaves of functions:

- Regular functions on a quasi-projective variety V form a sheaf $\mathcal{O}_V$ of $\mathbb{C}$-valued functions. The sheaf associates to each open set $U \subset V$ the $\mathbb{C}$-algebra $\mathcal{O}_V(U)$ of regular functions on U.

- Continuous $\mathbb{R}$-valued functions on a topological space form a sheaf of $\mathbb{R}$-valued functions.

- C^∞-functions on a smooth manifold form a sheaf of $\mathbb{R}$-valued functions.

- Holomorphic functions on a Riemann surface form a sheaf of $\mathbb{C}$-valued functions.

Definition: The sheaf $\mathcal{O}_V$ of regular functions on a quasi-projective variety V is called the *structure sheaf* of the variety.

The structure sheaf $\mathcal{O}_V$ determines V (up to isomorphism), even if we have only limited information about $\mathcal{O}_V$. For example, in Section 4.3, we proved that for an affine variety, the ring of global sections of $\mathcal{O}_V$ is the coordinate ring $\mathbb{C}[V]$, which in turn recovers the affine variety up to isomorphism. In other words, an affine variety is determined by the global sections of its structure sheaf. Only slightly more difficult is the fact that every quasi-projective variety is determined by the rings of sections of the structure sheaf on any affine cover, together with the restriction maps $\mathcal{O}_V(U_i) \to \mathcal{O}_V(U_i \cap U_j)$, which recover for us the way these affine pieces are glued together. Later in this appendix we will define an abstract variety, which will be determined by partial information about its structure sheaf in a similar way. This situation is unique to algebraic geometry: A manifold is not determined by such limited information about its sheaf of continuous (or differentiable, or complex holomorphic) functions.

For each open set $U \subset X$, a sheaf $\mathcal{R}$ has a natural *restriction* $\mathcal{R}|_U$ *to* U. The sections of $\mathcal{R}|_U$ over an open set $U' \subset U$ are just the sections $\mathcal{R}(U')$ of the original sheaf. Some caution is in order: The ring of sections $\mathcal{R}(U)$ and the restriction sheaf $\mathcal{R}|_U$ are two different objects; $\mathcal{R}(U)$ is a ring, while $\mathcal{R}|_U$ is a sheaf (assigning rings to open sets of U). For an open subset V of a quasi-projective variety W, the structure sheaf $\mathcal{O}_V$ agrees with the restriction of the sheaf $\mathcal{O}_W$ to the open set V.

A topological space, together with a sheaf of $\mathbb{C}$-valued functions on it, is an example of a *ringed space*. An understanding of ringed spaces is essential for eventually mastering the definition of a scheme, so we introduce the definition here.

Definition: A *sheaf of rings* $\mathcal{R}$ on a topological space X assigns to each open set $U \subset X$ a ring $\mathcal{R}(U)$ in such a way that the following axioms are satisfied

- If $U_1 \subset U_2$, then there is a ring homomorphism $\mathcal{R}(U_2) \to \mathcal{R}(U_1)$. This map is called "the restriction map from U_2 to U_1," and the image of any element $f \in \mathcal{R}(U_2)$ under this map is denoted by $f|_{U_1}$.

- If $U_1 \subset U_2 \subset U_3$, then the restriction map $\mathcal{R}(U_3) \to \mathcal{R}(U_1)$ is the composition of the restriction maps $\mathcal{R}(U_3) \to \mathcal{R}(U_2) \to \mathcal{R}(U_1)$.

- If $\{U_\alpha\}_{\alpha \in A}$ is an open cover of an open set $U \subset X$ and $\{g_\alpha\}_{\alpha \in A}$ is a collection of elements $g_\alpha \in \mathcal{R}(U_\alpha)$ such that for all indices α, β, $g_\alpha|_{U_\alpha \cap U_\beta} = g_\beta|_{U_\alpha \cap U_\beta}$, then there exists a unique $g \in \mathcal{R}(U)$ such that $g|_{U_\alpha} = g_\alpha$ for all α.

A topological space X, together with a sheaf of rings on X, is called a *ringed space*.

The rings $\mathcal{R}(U)$ in the definition above are just abstract rings: They need not be rings of functions on the set U. In particular, the word "restriction" above should not be interpreted literally as the restriction of functions.

Our sheaves of $\mathbb{C}$-valued functions are examples of sheaves of rings on the corresponding spaces. In fact, they are all sheaves of $\mathbb{C}$–algebras, since each ring $\mathcal{R}(U)$ is actually a $\mathbb{C}$–algebra. Although an abstract sheaf of rings is not necessarily a sheaf of functions, one should think of every sheaf of rings as very much like a sheaf of functions. The third axiom in the definition of a sheaf of rings—also called the sheaf axiom—guarantees that the elements of $\mathcal{R}(U)$ really behave like functions: They are uniquely defined by their values on any open cover of U. It is easy to check that every sheaf of $\mathbb{C}$-valued functions is a sheaf of rings.

We could also define a sheaf of Abelian groups, a sheaf of sets, a sheaf of algebras, or even a sheaf of objects in almost any category. Just strike out all occurrences of the word "ring" in the preceding definition and replace it by the word "Abelian group," "set," or "algebra."

A topological space may come equipped with several different sheaves of rings or algebras. For example, on $\mathbb{C}^n$ with its usual Euclidean topology, we have not only the sheaf of continuous functions, but also the sheaf of holomorphic functions. We can also equip the set $\mathbb{C}^n$ with the Zariski topology, where we have the sheaf of regular functions.

Definition: Let $\mathcal{R}$ and $\mathcal{S}$ be two sheaves of rings on a space X. A *map of sheaves of rings*

$$\mathcal{R} \xrightarrow{G} \mathcal{S}$$

consists of a ring map

$$\mathcal{R}(U) \xrightarrow{G(U)} \mathcal{S}(U)$$

for each open set $U \subset X$ such that whenever $U_1 \subset U_2$, the following diagram commutes:

$$
\begin{array}{ccc}
\mathcal{R}(U_2) & \xrightarrow{\;\;G(U_2)\;\;} & \mathcal{S}(U_2) \\
\downarrow & & \downarrow \\
\mathcal{R}(U_1) & \xrightarrow[\;\;G(U_1)\;\;]{} & \mathcal{S}(U_1)
\end{array}
$$

where the vertical maps are the restriction maps. If the sheaves of rings are actually sheaves of $\mathbb{C}$–algebras, the maps are furthermore required to preserve the $\mathbb{C}$–algebra structure, that is, each

$$\mathcal{R}(U) \xrightarrow{G(U)} \mathcal{S}(U)$$

must be $\mathbb{C}$-linear.

It does not make sense to speak of maps of sheaves when the sheaves are on two different topological spaces. However, given a continuous map $X \to Y$ of topological spaces, there is a way to define a sheaf of rings on Y from any sheaf of rings on X.

Definition: Given a sheaf $\mathcal{R}$ on a topological space X and a continuous map $X \xrightarrow{f} Y$ of topological spaces, the *push-forward* $f_*\mathcal{R}$ of $\mathcal{R}$ is the sheaf on Y defined as follows. For each open set $U \subset Y$,

$$f_*\mathcal{R}(U) = \mathcal{R}(f^{-1}(U)).$$

If $\mathcal{R}$ is a sheaf of rings on X, then $f_*\mathcal{R}$ is a sheaf of rings on Y.

Definition: A *map of ringed spaces* $(X, \mathcal{O}_X) \to (Y, \mathcal{O}_Y)$ is a pair $(F, F^{\#})$ consisting of a continuous map of topological spaces $X \xrightarrow{F} Y$ and a map of sheaves of rings on Y, $\mathcal{O}_Y \xrightarrow{F^{\#}} F_*\mathcal{O}_X$.

Example: Let $V \xrightarrow{F} W$ be a map of quasi-projective algebraic varieties. Then there is a naturally induced map of ringed spaces

$$(V, \mathcal{O}_V) \to (W, \mathcal{O}_W)$$

where $\mathcal{O}_V$ (respectively $\mathcal{O}_W$) is the sheaf of regular functions on V (respectively W). The map of topological spaces is F, and the map $\mathcal{O}_W \to F_*\mathcal{O}_V$

of sheaves of rings is defined by the pullback: For each open set $U \subset W$,

$$\mathcal{O}_W(U) \longrightarrow \mathcal{O}_V(F^{-1}(U)),$$
$$g \longmapsto F^{\#}(g) = g \circ F.$$

This idea works more generally, as the next example shows.

Example: If $X \xrightarrow{F} Y$ is any continuous map of topological spaces, there is always a morphism of ringed spaces

$$(X, \mathcal{F}_X) \to (Y, \mathcal{F}_Y),$$

where $\mathcal{F}_X$ is the sheaf of $\mathbb{C}$-valued functions on X and $\mathcal{F}_Y$ is the sheaf of $\mathbb{C}$-valued functions on Y. Indeed, the map of sheaves $\mathcal{F}_Y \to F_*\mathcal{F}_X$ is defined using the pullback,

$$\mathcal{F}_Y(U) \to \mathcal{F}_X(F^{-1}(U)),$$
$$g \longmapsto g \circ F.$$

If $\mathcal{F}_X$ and $\mathcal{F}_Y$ instead denote the sheaves of *continuous* $\mathbb{C}$-valued functions on X and Y, then $(X, \mathcal{F}_X) \to (Y, \mathcal{F}_Y)$ will be a morphism of these ringed spaces. More generally, if X and Y have some more refined structure of ringed spaces via sheaves of functions $\mathcal{O}_X$ and $\mathcal{O}_Y$, it is often possible to define a map of ringed spaces in the same way. There is always a pullback map

$$\mathcal{O}_Y(U) \to \mathcal{F}_X(F^{-1}(U)),$$

where $\mathcal{F}_X$ is the sheaf of all $\mathbb{C}$-valued functions on X. One must check only that the pullback of a function in $\mathcal{O}_Y(U)$ lies in the subring of functions $\mathcal{O}_X(F^{-1}(U)) \subseteq \mathcal{F}_X(F^{-1}(U))$. For instance, if X and Y are smooth manifolds and $\mathcal{O}_X$ and $\mathcal{O}_Y$ are the corresponding sheaves of smooth functions on X and Y, then any *smooth* map $X \xrightarrow{F} Y$ induces a morphism of ringed spaces:

$$(X, \mathcal{O}_X) \xrightarrow{(F, F^{\#})} (Y, \mathcal{O}_Y).$$

Here, $F^{\#}$ is completely determined by F in a natural way. In dealing with abstract ringed spaces, where the sheaf of rings may not be a sheaf of functions on X, things sometimes get more complicated. This more abstract point of view is necessary, however, in laying the foundations of scheme theory (see, for instance, [20, Chapter II, Sections 1 and 2]).

Definition: A morphism of ringed spaces $(X, \mathcal{O}_X) \xrightarrow{(F, F^{\#})} (Y, \mathcal{O}_Y)$ is an *isomorphism* if it has an inverse. More precisely, we require the existence of a morphism of ringed spaces

$$(Y, \mathcal{O}_Y) \xrightarrow{(G, G^{\#})} (X, \mathcal{O}_X)$$

such that $X \xrightarrow{F} Y \xrightarrow{G} X$ is the identity on X and

$$\mathcal{O}_X \xrightarrow{G^{\#}} G_*\mathcal{O}_Y \xrightarrow{F^{\#}} (G \circ F)_*\mathcal{O}_X = \mathcal{O}_X$$

is the identity map of sheaves, and similarly $Y \xrightarrow{G} X \xrightarrow{F} Y$ and

$$\mathcal{O}_Y \xrightarrow{F^{\#}} F_*\mathcal{O}_Y \xrightarrow{G^{\#}} (F \circ G)_*\mathcal{O}_Y = \mathcal{O}_Y$$

are identity maps.

Working with morphisms of ringed spaces requires a considerable amount of notation, but is really quite natural and becomes easy with practice. A detailed exposition for beginners struggling with the notation can be found on the web at

http://www.math.lsa.umich.edu/~kesmith/inverse.ps.

A.2   Abstract Algebraic Varieties

An *abstract algebraic variety* is a topological space that has an open cover by sets that are homeomorphic to affine algebraic varieties—possibly in ambient affine spaces of different dimensions—glued together by transition functions that are morphisms of affine algebraic varieties. The easiest way to make this precise is to use the sheaf of regular functions of an affine variety.

Definition: A *complex abstract algebraic variety* is a ringed space $(V, \mathcal{O}_V)$ that has an open cover $V = \bigcup U_\lambda$ where each $(U_\lambda, \mathcal{O}_V|_{U_\lambda})$ is isomorphic as a ringed space to some affine algebraic variety $(W_\lambda, \mathcal{O}_{W_\lambda})$, together with its structure sheaf $\mathcal{O}_{W_\lambda}$.

Explicitly, each U_λ admits a homeomorphism $U_\lambda \xrightarrow{H_\lambda} W_\lambda$ with an affine variety W_λ such that the pullback mapping $H_\lambda^{\#}$ induces an isomorphism

$$\mathcal{O}_{W_\lambda} \xrightarrow{H_\lambda^{\#}} H_{\lambda*}\mathcal{O}_{U_\lambda}$$

of sheaves of $\mathbb{C}$-valued functions on W_λ. That is, for each open set $U \subset W_\lambda$, the map

$$\mathcal{O}_{W_\lambda}(U) \xrightarrow{H_\lambda^{\#}(U)} H_{\lambda*}\mathcal{O}_{U_\lambda}(U) = \mathcal{O}_{U_\lambda}(H_\lambda^{-1}(U)),$$
$$g \longmapsto g \circ H_\lambda,$$

is an isomorphism of $\mathbb{C}$–algebras. Of course, we may replace $\mathbb{C}$ here by any algebraically closed field k to define an abstract algebraic variety over k.

The sheaf $\mathcal{O}_V$ is called the *structure sheaf* of the variety V, and its sections over an open set U are called *regular functions over U*. The definition of an abstract variety is similar to the definition of abstract geometric objects in other categories. For example, a smooth manifold can be defined as a ringed space $(M, \mathcal{C}^\infty)$ that has an open cover $\bigcup U_\lambda$ such that $(U_\lambda, \mathcal{C}^\infty|_{U_\lambda})$ is isomorphic as a ringed space to $(B, \mathcal{C}^\infty_B)$, where $B \subset \mathbb{R}^n$ is an open ball and $\mathcal{C}^\infty_B$ is the sheaf of smooth functions on B. Likewise, a complex manifold can be defined as a ringed space $(M, \mathcal{H})$ that has an open cover $\bigcup U_\lambda$ such that each $(U_\lambda, \mathcal{H}|_{U_\lambda})$ is isomorphic as a ringed space to $(B, \mathcal{H}_B)$, where $B \subset \mathbb{C}^n$ is a complex open ball and $\mathcal{H}_B$ is the sheaf of holomorphic functions on B.

A *morphism of abstract varieties* $(V, \mathcal{O}_V) \to (W, \mathcal{O}_W)$ is simply a morphism between the corresponding ringed spaces that also preserves the $\mathbb{C}$-algebra structure. That is, it is a morphism of "$\mathbb{C}$-algebra-ed spaces," meaning that for each open set U of W, the corresponding map $\mathcal{O}_W(U) \to \mathcal{O}_V(F^{-1}(U))$ is a homomorphism of $\mathbb{C}$-algebras, not just a map of rings. An *isomorphism of abstract varieties* is defined in the obvious way.

Quasi-projective varieties, together with their structure sheaves, are examples of abstract algebraic varieties. These form a large class of interesting objects, and they are the only varieties considered by many algebraic geometers. Abstract algebraic varieties arise naturally in the study of quasi-projective (or even affine or projective) varieties. For example, they appear as moduli spaces of quasi-projective varieties, such as the moduli space $\mathfrak{M}_g$ of genus-g projective curves mentioned in Sections 7.6 and A.1. In these cases it may be useful to know that the abstractly defined variety is in fact quasi-projective, but often it is not so important whether or not the variety is quasi-projective.

Usually, an additional property, called separatedness, is included as part of the definition of an abstract algebraic variety. A complex abstract algebraic variety as defined above is *separated* if it is Hausdorff in the Euclidean topology. For varieties defined over fields other than $\mathbb{C}$, the definition of separatedness is somewhat more technical (see [20, page 95]). All quasi-projective varieties are separated. An example of a non-separated variety is a line with a doubled origin, also known as the "bug-eyed line": Two copies of $\mathbb{A}^1$ identified at all points except at 0. See [37, Chapter V, page 44].

As we have seen, the spectrum (or at least the collection of closed points in the spectrum) of a finitely generated reduced $\mathbb{C}$-algebra can be identified with an affine algebraic variety. Abstract varieties are just ringed spaces admitting an open cover whose associated rings $\mathcal{R}(U)$ are all finitely generated reduced $\mathbb{C}$-algebras. By relaxing these conditions on the rings $\mathcal{R}(U)$, for instance by allowing $\mathcal{R}(U)$ to have nilpotents or even dropping the restriction that it be a $\mathbb{C}$-algebra, we arrive at the definition of a scheme.

A *scheme* is a natural generalization of an abstract algebraic variety. A scheme is also defined as a ringed space, but the open sets in the cover are modeled on affine schemes, instead of on affine algebraic varieties. We earlier defined an affine scheme as the prime spectrum $\mathrm{Spec}(R)$ of some ring R, considered as a topological space with its Zariski topology. There is a natural way to define a sheaf of rings $\tilde{R}$ on the topological space $\mathrm{Spec}(R)$, in such a way that the global sections of this sheaf recover R. (This requires some slightly technical algebra, so we do not go into this here.) So the rough definition of a scheme can be stated as follows: A *scheme* is a ringed space $(X, \mathcal{O}_X)$ that admits an open cover $\bigcup U_\lambda$ such that each $(U_\lambda, \mathcal{O}_X|_{U_\lambda})$ is isomorphic as a ringed space to some affine scheme $\mathrm{Spec} R_\lambda$ with its natural sheaf of rings $\tilde{R}_\lambda$. The rings R_λ may be completely arbitrary: They need not be rings of functions or any kind of finitely generated reduced $\mathbb{C}$-algebras, as would be the case for algebraic varieties. To make this definition precise, we would actually need to define the notion of a *locally ringed space,* which requires that if we take a limit over all open sets containing a given point of the scheme, the corresponding limit of rings is a so-called *local ring.* Rather than go into this here, we refer the reader to any textbook on the subject.

The theory of schemes is a beautiful subject, fundamental to modern algebraic geometry. For the basic theory of schemes, the reader should consult [37, Chapter V], [20, Chapter 2], or [10].

References

[1] Abramovich, D. and de Jong, A. J. *Smoothness, semistability, and toroidal geometry.* J. Algebraic Geom. **6** 1997, no. 4, 789–801.

[2] Beauville, Arnaud. *Complex algebraic surfaces.* Translated from the 1978 French original by R. Barlow, with assistance from N. I. Shepherd-Barron and M. Reid. Second edition. London Mathematical Society Student Texts **34**. Cambridge University Press, Cambridge, 1996.

[3] Biersone, Edward and Milman, Pierre D. *Canonical desingularization in characteristic zero by blowing up the maximal strata of a local invariant.* Invent. Math. **128** 1997, #2, 207–302. Reviewed in Math Reviews, 98e:14010.

[4] Bogomolov, Fedor A. and Pantev, Tony G. *Weak Hironaka theorem.* Math. Res. Lett. **3** 1996, no. 3, 299–307.

[5] Cox, David and Little, John and O'Shea, Donal. *Ideals, Varieties, and Algorithms.* Undergraduate Texts in Mathematics. Springer-Verlag, 1992.

[6] Cox, David and Little, John and O'Shea, Donal. *Using Algebraic Geometry.* Graduate Texts in Mathematics **185**. Springer-Verlag, 1998.

[7] de Jong, A. J. *Smoothness, Semi-stability and Alterations.* Inst. Hautes Études Sci. Publ. Math. **83** 1996, 51–93.

[8] Deligne, P. and Mumford, D. *The irreducibility of the space of curves of given genus.* Inst. Hautes Études Sci. Publ. Math. **36** 1969, 75–109.

[9] Eisenbud, David. *Commutative Algebra with a View Toward Algebraic Geometry.* Graduate Texts in Mathematics **150**. Springer-Verlag, 1995.

[10] Eisenbud, David and Harris, Joe. *The Geometry of Schemes.* Graduate Texts in Mathematics **197**. Springer-Verlag, 2000.

[11] Fulton, William. *Introduction to intersection theory in algebraic geometry.* CBMS Regional Conference Series in Mathematics **54**. American Mathematical Society, 1984.

[12] Fulton, William. *Intersection Theory.* Second Edition. Springer-Verlag, 1998.

[13] Fulton, William. *Young Tableau. With applications to representation theory and geometry.* London Mathematical Society Student Texts, **35**. Cambridge University Press, 1997.

[14] Fulton, William. *Algebraic curves. An introduction to algebraic geometry.* Notes written with the collaboration of Richard Weiss. Addison-Wesley Publishing Company, 1989.

[15] Griffiths, Phillip and Harris, Joe. *Principles of Algebraic Geometry.* John Wiley and Sons, New York, 1978.

[16] Mumford, D. and Fogarty, J. and Kirwan, F. *Geometric invariant theory.* Third edition. Springer-Verlag, 1994.

[17] Harris, Joe. *Algebraic Geometry. A First Course.* Gradute Texts in Mathematics **133**. Springer-Verlag, 1992.

[18] Harris, Joe. *An introduction to the moduli space of curves.* Mathematical Aspects of String Theory (San Diego, CA 1986), 285–312, Adv. Ser. Math. Phys. **1**, World Sci., 1987.

[19] Harris, Joe and Morrison, Ian. *Moduli of curves.* Graduate Texts in Mathematics **187**. Springer-Verlag, 1998.

[20] Hartshorne, Robin. *Algebraic Geometry.* Graduate Texts in Mathematics **52**. Springer-Verlag, 1977.

[21] Hironaka, Heisuke. *Resolution of singularities of an algebraic variety over a field of characteristic zero. I, II.* Ann. of Math **70** 1964, 109–203; **79** 1964, 205–326.

[22] Hilbert, David. *Über die Theorie von algebraischen Formen.* Math. Ann. **36** 1890, 473–534.

[23] Hilbert, David. *Theory of algebraic invariants.* Translated from the German and with a preface by Reinhard C. Laubenbacher. Edited and with an introduction by Bernd Sturmfels. Cambridge University Press, Cambridge, 1993.

[24] Hoffman, Kenneth and Ray Kunze. *Linear Algebra.* Second Edition. Prentice–Hall, 1971.

[25] Kleiman, S. L. and Dan Laksov. *Schubert Calculus.* Amer. Math. Monthly **79** 1972, 1061–1082.

[26] Kollár, János. *Sharp Effective Nullstellensatz.* J. Amer. Math. Soc. **1** 1988, #4, 963–975.

[27] Kollár, János. *The structure of algebraic threefolds: an introduction to Mori's program.* Bull. Amer. Math. Soc. (N.S.) **17** 1987, #2, 211–273.

[28] Kollár, János. *Real Algebraic Surfaces.* Princeton University Preprint, 2000.

[29] Kollár, János. *Singularities of Pairs,* in Proceedings of the 1995 conference in Algebraic Geometry, Santa Cruz, American Mathematical Society Symposia, 1997.

[30] Lazarsfeld, Robert. *Lectures on Linear Series.* With the assistance of Guilleno Fernández del Busto. IAS/Park City Math Ser. **3**, Complex Algebraic Geometry (Park City, UT 1993), 161–219. Amer. Math. Soc., 1997.

[31] Lipman, J. Review of *Canonical desingularization in characteristic zero by blowing up the maximum strata of a local invariant* by E. Bierstone and P. Milman. Review 98e:14010, Mathematical Reviews, 1998.

[32] Miranda, Rick. *Algebraic Curves and Riemann Surfaces.* American Mathematical Society, Graduate Studies in Mathematics **5**, 1995.

[33] Mumford, David. *Lectures on Curves on an Algebraic Surface.* With a section by G. M. Bergman. Annals of Mathematics Study **59**. Princeton University Press, 1966.

[34] Reid, Miles. *Chapters on Algebraic Surfaces.* IAS/Park Cirt Math Ser. **3**, Complex Algebraic Geometry (Park City, UT 1993), 3–159. Amer. Math. Soc., 1997.

[35] Reid, Constance. *Hilbert.* Reprint of the 1970 original. Copernicus, New York, 1996.

[36] Serre, Jean–Pierre. *Géometrie algébraique et géometrie analytique (French).* Ann. Inst. Fourier, Grenoble. **6** 1955-1956, 1–42.

[37] Shafarevich, Igor R. *Basic Algebraic Geoemtry.* First Edition. Grundlehren **213**, Springer-Verlag, 1974.

[38] Silhol, Robert. *Real algebraic surfaces.* Lecture Notes in Mathematics **1392**. Springer-Verlag, 1989.

[39] Villamayor, Orlando. *Constructiveness of Hironaka's Resolution.* Ann. Sci. École Norm. Sup. **22** 1989, #1, 1–32.

Index

Universitext   *(continued)*

Jones/Morris/Pearson: Abstract Algebra and Famous Impossibilities
Kac/Cheung: Quantum Calculus
Kannan/Krueger: Advanced Analysis
Kelly/Matthews: The Non-Euclidean Hyperbolic Plane
Kostrikin: Introduction to Algebra
Kurzweil/Stellmacher: The Theory of Finite Groups: An Introduction
Luecking/Rubel: Complex Analysis: A Functional Analysis Approach
MacLane/Moerdijk: Sheaves in Geometry and Logic
Marcus: Number Fields
Martinez: An Introduction to Semiclassical and Microlocal Analysis
Matsuki: Introduction to the Mori Program
McCarthy: Introduction to Arithmetical Functions
McCrimmon: A Taste of Jordan Algebras
Meyer: Essential Mathematics for Applied Fields
Mines/Richman/Ruitenburg: A Course in Constructive Algebra
Moise: Introductory Problems Course in Analysis and Topology
Morris: Introduction to Game Theory
Poizat: A Course In Model Theory: An Introduction to Contemporary Mathematical Logic
Polster: A Geometrical Picture Book
Porter/Woods: Extensions and Absolutes of Hausdorff Spaces
Radjavi/Rosenthal: Simultaneous Triangularization
Ramsay/Richtmyer: Introduction to Hyperbolic Geometry
Reisel: Elementary Theory of Metric Spaces
Ribenboim: Classical Theory of Algebraic Numbers
Rickart: Natural Function Algebras
Rotman: Galois Theory
Rubel/Colliander: Entire and Meromorphic Functions
Sagan: Space-Filling Curves
Samelson: Notes on Lie Algebras
Schiff: Normal Families
Shapiro: Composition Operators and Classical Function Theory
Simonnet: Measures and Probability
Smith: Power Series From a Computational Point of View
Smith/Kahanpää/Kekäläinen/Traves: An Invitation to Algebraic Geometry
Smorynski: Self-Reference and Modal Logic
Stillwell: Geometry of Surfaces
Stroock: An Introduction to the Theory of Large Deviations
Sunder: An Invitation to von Neumann Algebras
Tondeur: Foliations on Riemannian Manifolds
Toth: Finite Möbius Groups, Minimal Immersions of Spheres, and Moduli
van Brunt: The Calculus of Variations
Wong: Weyl Transforms
Zhang: Matrix Theory: Basic Results and Techniques
Zong: Sphere Packings
Zong: Strange Phenomena in Convex and Discrete Geometry